GLENCOE MATH

Interactive Guide

CONTRIBUTING AUTHORS

Philip Gonsalves
Director of Curriculum and Instruction for Mathematics
West Contra Costa Unified School District
Richmond, California

Dinah Zike
Educational Consultant
Dinah-Might Activities, Inc.
San Antonio, Texas

Mc
Graw
Hill
Education

Bothell, WA • Chicago, IL • Columbus, OH • New York, NY

mheducation.com/prek-12

STEM McGraw-Hill is committed to providing
instructional materials in Science, Technology, Engineering,
and Mathematics (STEM) that give all students a solid
foundation, one that prepares them for college and careers
in the 21st century.

Send all inquiries to:
McGraw-Hill Education
STEM Learning Solutions Center
8787 Orion Place
Columbus, OH 43240

ISBN: 978-0-07-898942-1
MHID: 0-07-898942-6

Printed in the United States of America.

1 2 3 4 5 6 7 8 9 LWI 22 21 20 19 18 17

Visual Kinesthetic Vocabulary® is a registered trademark of
Dinah-Might Adventures, LP.

Contents

Chapter 5 Expressions

Chapter 6 Equations and Inequalities

Chapter 7 Geometric Figures

Chapter 8 Measure Figures

Chapter 9 Probability

Chapter 10 Statistics

Inquiry Lab Guided Writing

Unit Rates

HOW can you use a bar diagram to solve a real-world problem involving ratios?

Use the exercises below to help answer the Inquiry Question.
Write the correct word or phrase on the lines provided.

1. Rewrite the question in your own words.

_____

_____

2. What key words do you see in the question?

_____

3. A _____ shows the relationship between two quantities. An example is 5:2.

4. If you make a bar diagram to represent the ratio 5:2, how many bars should you draw? _____

5. Into how many parts would you divide the first bar? _____

6. Into how many parts would you divide the second bar? _____

Use the bar diagram below to answer Exercises 7 – 10.

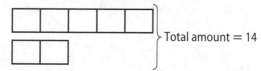

7. The ratio of dogs to cats in a neighborhood is 5 to 2. What does the top bar represent? _____

8. What does the bottom bar represent? _____

9. Complete the equation represented by the bar diagram.

_____ $x +$ _____ $x \ = \ 14$

HOW can you use a bar diagram to solve a real-world problem involving ratios?

_____

_____

_____

Lesson 1 Vocabulary

Rates

Use the word cards to define each vocabulary word or phrase and give an example.

Word Cards

rate	tasa

Definition

_____

_____

_____

Definición

_____

_____

_____

Example Sentence

_____

_____

- -

Word Cards

unit rate	tasa unitaria

Definition

_____

_____

_____

Definición

_____

_____

_____

Example Sentence

_____

_____

Lesson 2 Vocabulary

Complex Fractions and Unit Rates

Use the definition map to list qualities about the vocabulary word or phrase.

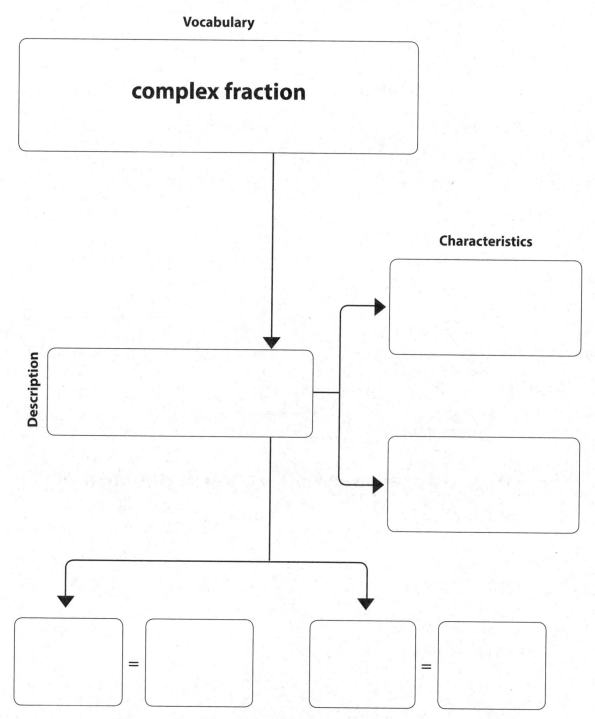

Vocabulary

complex fraction

Characteristics

Description

Write and simplify examples of complex fractions.

Lesson 3 Vocabulary

Convert Unit Rates

Use the word cards to define each vocabulary word or phrase and give an example.

Word Cards

unit ratio	razón unitaria
Definition	**Definición**
_____	_____
_____	_____
_____	_____
Example Sentence	
_____	
_____	

- -

Word Cards

dimensional analysis	análisis dimensional
Definition	**Definición**
_____	_____
_____	_____
_____	_____
Example Sentence	
_____	
_____	

Lesson 4 Vocabulary

Proportional and Nonproportional Relationships

Use the three column chart to organize the vocabulary in this lesson. Write the word in Spanish. Then write the definition of each word.

English	Spanish	Definition
unit ratio		
rate		
proportional		
nonproportional		
equivalent ratios		

Problem-Solving Investigation
The Four-Step Plan

Case 3 Financial Literacy

Terry opened a savings account in **December** with **$150** and deposited **$30 each month** beginning in **January**.

What is the **value** of Terry's account at the **end of July**?

- Understand:

- Plan:

- Solve:

- Check:

Case 4 STEM

About **how many centimeters longer** is the average **femur** than the average **tibia**? (Hint 1 inch ≈ 2.54 centimeters)

- Understand:

- Plan:

- Solve:

- Check:

Bones in a Human Leg	
Bone	**Length (in.)**
Femur (upper leg)	19.88
Tibia (inner lower leg)	16.94
Fibula (outer lower leg)	15.94

Lesson 5 Vocabulary

Graph Proportional Relationships

Use the word bank to identify the parts of the coordinate plane. Then draw an arrow from the word to the part of the coordinate plane it describes.

Word Bank			
origin	ordered pair	Quadrant III	Quadrant I
x-axis	x-coordinate	y-axis	y-coordinate

_____ _____

_____   _____

(−3, 2)

(2, 1)

_____

_____

_____

_____

Inquiry Lab Guided Writing

Proportional and Nonproportional Relationships

HOW are proportional and nonproportional linear relationships alike? HOW are they different?

Use the exercises below to help answer the Inquiry Question. Write the correct word or phrase on the lines provided.

1. Rewrite the question in your own words.

   _____

   _____

2. What key words do you see in the question?

   _____

3. The graph of a _____ relationship is a straight line.

4. Two linear relationships that have a constant rate are _____ .

5. Two linear relationships with a rate that is not constant are _____ .

6. Is the graph of a proportional linear relationship a straight line? _____

7. What does the graph of a nonproportional linear relationship look like?

   _____

8. What is the origin on a coordinate plane?

   _____

9. For which type of linear relationship does the graph pass through the origin?

   _____

   HOW are proportional and nonproportional linear relationships alike?
   HOW are they different?

   _____

   _____

   _____

Lesson 6 Vocabulary
Solve Proportional Relationships

Use the word cards to define each vocabulary word or phrase
and give an example.

Word Cards

proportion	proporción
Definition	**Definición**
_____	_____
_____	_____
_____	_____
Example Sentence	
_____	
_____	

- -

Word Cards

cross product	producto cruzado
Definition	**Definición**
_____	_____
_____	_____
_____	_____
Example Sentence	
_____	
_____	

Inquiry Lab Guided Writing

Rate of Change

HOW is unit rate related to rate of change?

Use the exercises below to help answer the Inquiry Question. Write the correct word or phrase on the lines provided.

1. Rewrite the question in your own words.

_____

_____

2. What key words do you see in the question?

_____

_____

3. A rate with a denominator of 1 unit is called a _____ .

4. A _____ describes how one quantity changes in relation to another.

Use the table below to answer Exercises 5-8.

Bike Rental	
Number of Days	**Cost ($)**
1	8
2	16
3	24
4	32

5. The unit rate is $8 per _____ .

6. As the number of days increase by 1, the cost increases by _____ .

7. Write the rate of change as a fraction. _____

8. Are the unit rate and the rate of change the same? _____

HOW is unit rate related to rate of change?

_____

_____

Lesson 7 Vocabulary

Constant Rate of Change

Use the definition map to list qualities about the vocabulary word or phrase.

Vocabulary

rate of change

**Characteristics:
What it is.**

Description

Examples

Lesson 8 Vocabulary
Slope

Use the four-square chart to define slope in different ways.

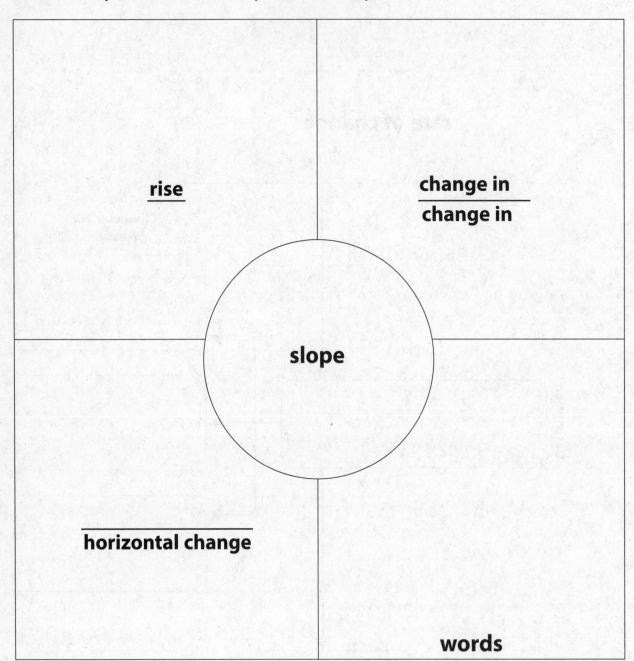

rise
_____

change in
_____
change in

slope

_____
horizontal change

words

Name two ways you can find the slope of a line.

_____

_____

Lesson 9 Vocabulary

Direct Variation

Use the three column chart to organize the vocabulary in this lesson. Write the word in Spanish. Then write the definition of each word.

English	Spanish	Definition
direct variation		
constant of variation		
constant of proportionality		
unit rate		

Inquiry Lab Guided Writing

Percent Diagrams

HOW are percent diagrams used to solve real-world problems?

Use the exercises below to help answer the Inquiry Question.
Write the correct word or phrase on the lines provided.

1. Rewrite the question in your own words.

    _____

    _____

2. What key words do you see in the question?

    _____

3. What type of model, or diagram, is easy to divide into equal parts? _____

4. You can use a bar diagram to show percents. What value would the full

    length of the bar represent? _____

Use the bar diagrams below to answer Exercises 5 - 10. Suppose you want to
find 30% of 200 teachers. Label what each bar diagram shows.

5. _____ | 10% | 10% | 10% | 10% | 10% | 10% | 10% | 10% | 10% | 10% | 100%

6. _____ | 20 | 20 | 20 | 20 | 20 | 20 | 20 | 20 | 20 | 20 | _____ total teachers

7. How many bar diagrams are needed to model the percent of a number? _____

8. Are the bar diagrams the same length or different lengths? _____

9. Are both bar diagrams divided into equal parts? _____

10. How many sections of each bar diagram would you shade to find

    30% of 200 teachers? _____

HOW are percent diagrams used to solve real-world problems?

_____

_____

_____

Lesson 1 Notetaking

Percent of a Number

Use Cornell notes to better understand the lesson's concepts. Complete each sentence by filling in the blanks with the correct word or phrase.

Questions	Notes
1. How do I find the percent of a number?	I can write the percent as a _____ or a _____ and then _____ .
2. How do I use a percent greater than 100%?	I can write the percent as an _____ fraction, _____ , or a _____ greater than _____ .

Summary
Give an example of a real-world situation in which you would find the percent of a number.
_____
_____
_____
_____
_____
_____
_____

Lesson 2 Review Vocabulary

Percent and Estimation

Use the flow chart to review the processes for estimating the percent of a number.

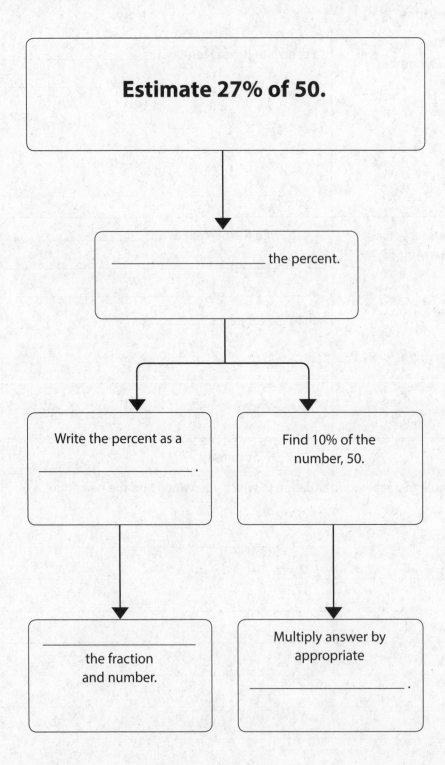

Estimate 27% of 50.

_____ the percent.

Write the percent as a
_____ .

Find 10% of the
number, 50.

_____
the fraction
and number.

Multiply answer by
appropriate
_____ .

Inquiry Lab Guided Writing

Find Percents

HOW is percent used to solve real-world problems?

Use the exercises below to help answer the Inquiry Question.
Write the correct word or phrase on the lines provided.

1. Rewrite the question in your own words.

_____

_____

2. What key words do you see in the question?

_____

Use the information and bar diagrams below to answer Exercises 3 - 6.

Ace Candy Company makes 50 kinds of candy. 30 of the candies have chocolate.
Zip Candy Company makes 70 kinds of candy. 28 of the candies have chocolate.

Ace ├ ------ 50 kinds of candy ------ ┤
| 5 | 5 | 5 | 5 | 5 | 5 | 5 | 5 | 5 | 5 | 100%

Zip ├ ------ 70 kinds of candy ------ ┤
| 7 | 7 | 7 | 7 | 7 | 7 | 7 | 7 | 7 | 7 | 100%

3. What percent does each section of the bar diagrams represent? _____

4. How many parts of the top bar diagram show the percentage of Ace candies

with chocolate? _____ Shade that number of parts.

5. How many parts of the bottom bar diagram show the percentage of Zip candies

with chocolate? _____ Shade that number of parts.

6. Use the bar diagrams to compare the two companies. Which company makes a

greater percentage of candies with chocolate? _____

HOW is percent used to solve real-world problems?

_____

_____

_____

Lesson 3 Vocabulary

The Percent Proportion

Use the definition map to list qualities about the vocabulary word or phrase.

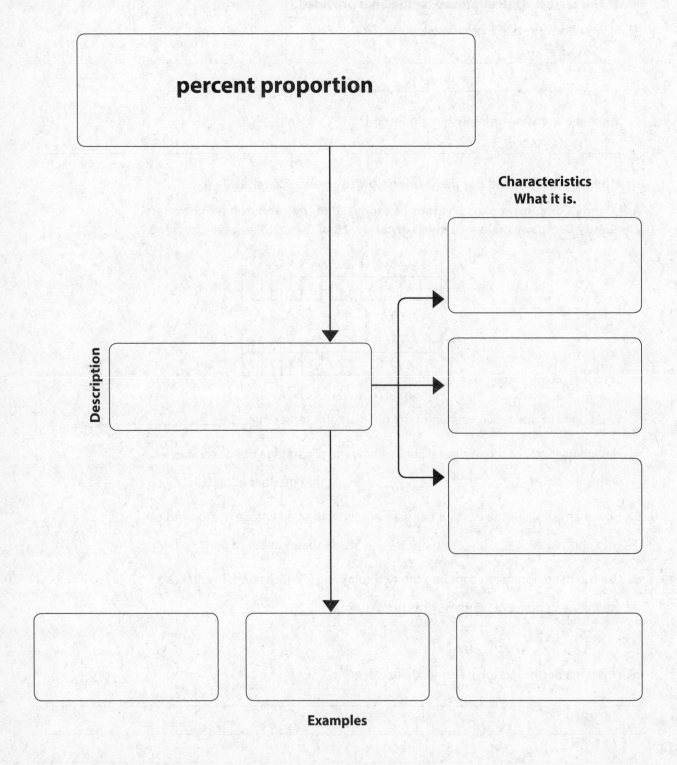

percent proportion

Characteristics
What it is.

Description

Examples

Lesson 4 Review Vocabulary

The Percent Equation

Use the concept web to show the parts of a percent equation. Use the words part, percent and whole.

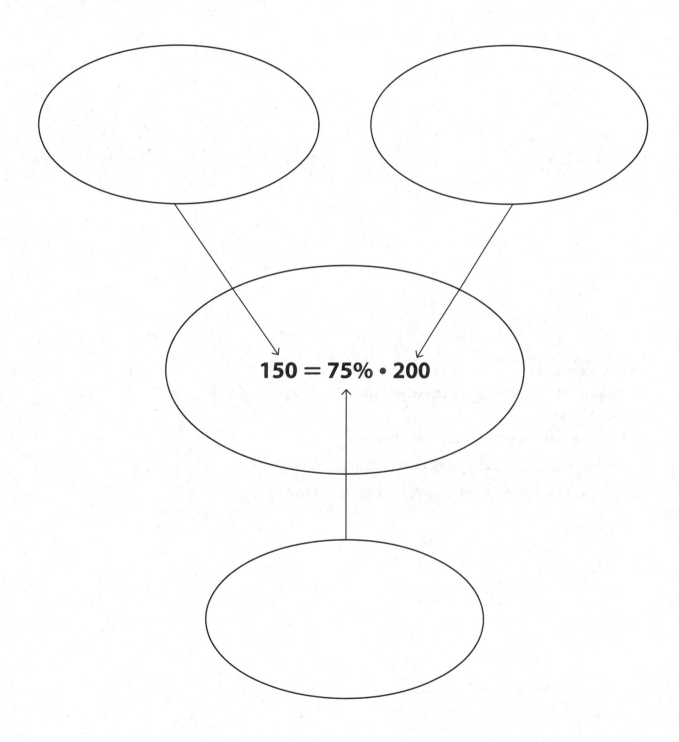

$$150 = 75\% \cdot 200$$

Problem-Solving Investigation
Determine Reasonable Answers

Case 3 Travel

A travel agency surveyed **140** families about their favorite vacation spots.

Is it **reasonable** to say that **24 more** families chose **Hawaii** over **Florida**? Explain.

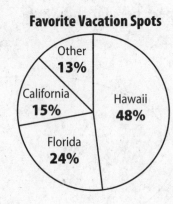

Favorite Vacation Spots

Other 13%
California 15%
Hawaii 48%
Florida 24%

- Understand:

- Plan:

- Solve:

- Check:

Case 4 Exercise

A survey showed that **61% of middle school students** do some kind of physical activity every day.

Of those students, **9% play on the football team**.

Suppose there are **828 middle school students** in your school.

About **how many students** would play on the **football team**?

- Understand:

- Plan:

- Solve:

- Check:

Inquiry Lab Guided Writing

Percent of Change

HOW can you use a bar diagram to show a percent of change?

**Use the exercises below to help answer the Inquiry Question.
Write the correct word or phrase on the lines provided.**

1. Rewrite the question in your own words.

_____

_____

2. What key words do you see in the question?

_____

3. The full length of a bar diagram represents what percent? _____

4. If you want to show an increase to the original quantity, how could you change

the bar diagram? _____

5. If you want to show a decrease from the original amount, how could you change

the bar diagram?_____

Use the bar diagram below to answer Exercises 6 - 9.

6. What number represents 100%? _____

7. Does the bar diagram show an increase or decrease from Lunches Sold Last Week

to Lunches Sold This Week? _____

8. What is the percent of change? _____

9. How could you change the bar diagram to show an increase of 30%?

_____

HOW can you use a bar diagram to show a percent of change?

_____

_____

_____

Lesson 5 Vocabulary

Percent of Change

Use the vocabulary squares to write a definition, a sentence, and an example for each vocabulary word.

	Definition
percent of change	
Example	**Sentence**

	Definition
percent of increase	
Example	**Sentence**

	Definition
percent of decrease	
Example	**Sentence**

Lesson 6 Vocabulary

Sales Tax, Tips, and Markup

Use the three column chart to organize the vocabulary in this lesson. Write the word in Spanish. Then write the definition of each word.

English	Spanish	Definition
sales tax		
tip		
gratuity		
markup		
selling price		

Lesson 7 Vocabulary

Discount

Use the word cards to define each vocabulary word or phrase and give an example.

Word Cards

discount	**descuento**
Definition	**Definición**
_____	_____
_____	_____
_____	_____
Example Sentence	
_____	_____
_____	_____

Word Cards

markdown	**rebaja**
Definition	**Definición**
_____	_____
_____	_____
_____	_____
Example Sentence	
_____	_____
_____	_____

Lesson 8 Vocabulary

Financial Literacy: Simple Interest

Use the concept web to identify and describe the parts of the simple interest formula.

Word Bank			
simple interest	principal	annual interest rate	time

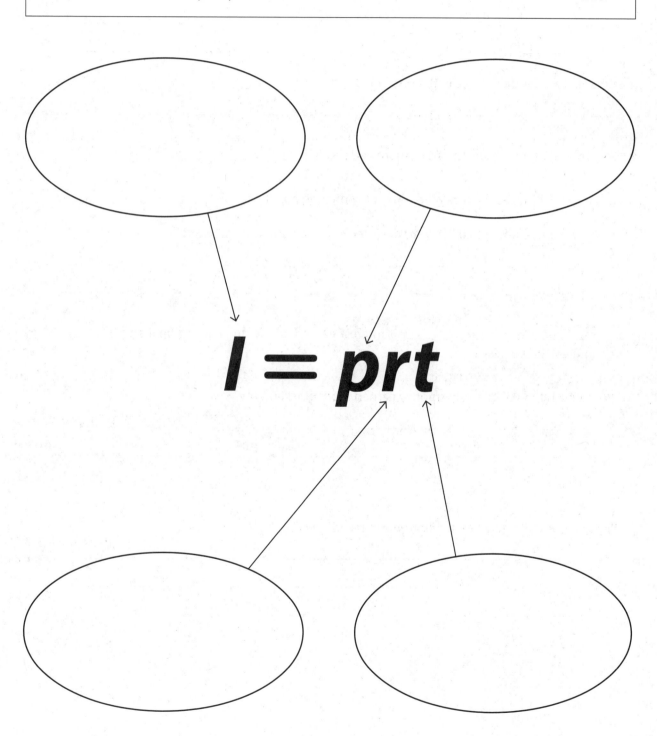

Inquiry Lab Guided Writing

Spreadsheet: Compound Interest

HOW is compound interest different from simple interest?

Use the exercises below to help answer the Inquiry Question.
Write the correct word or phrase on the lines provided.

1. Rewrite the question in your own words.

_____

_____

2. What key words do you see in the question?

_____

3. An amount of money deposited or borrowed is called _____ .

4. What type of interest is calculated using only the original principal? _____

5. Write the formula for calculating simple interest using variables and words.

_____

_____

6. _____ is interest earned on the original principal and
on interest already earned.

7. With which type of interest is money earned more quickly? Why?

_____

_____

HOW is compound interest different from simple interest?

_____

_____

_____

Lesson 1 Vocabulary

Integers and Absolute Value

Use the three column chart to organize the vocabulary in this lesson. Write the word in Spanish. Then write the definition of each word.

English	Spanish	Definition
integer		
negative integer		
positive integer		
graph		
absolute value		

Inquiry Lab Guided Writing

Add Integers

WHEN is the sum of two integers a negative number?

**Use the exercises below to help answer the Inquiry Question.
Write the correct word or phrase on the lines provided.**

1. Rewrite the question in your own words.

_____

_____

2. What key words do you see in the question?

_____

3. A _____ is the answer to an addition problem.

4. Any positive or negative whole number, including zero, is an _____ .

5. Negative integers are _____ than zero, and positive

integers are _____ than zero.

6. The number of units an integer is from 0 on a number line is its _____ .

Use the following equation to answer Exercises 7 and 8: $-7 + (-12) = -19$.

7. Are both addends negative or positive? _____

8. Is the sum negative or positive? _____

Use the following equation to answer Exercises 9 and 10 : $-20 + 4 = -16$.

9. Which addend has the greater absolute value? _____

10. Is the sum negative or positive? _____

WHEN is the sum of two integers a negative number?

_____

_____

_____

_____

Lesson 2 Vocabulary

Add Integers

Use the word cards to define each vocabulary word or phrase and give an example.

Word Cards

opposites	**opuestos**
Definition	**Definición**
_____	_____
_____	_____
_____	_____
Example Sentence	
_____	
_____	

Word Cards

additive inverse	**inverso aditivo**
Definition	**Definición**
_____	_____
_____	_____
_____	_____
Example Sentence	
_____	
_____	

Inquiry Lab Guided Writing

Subtract Integers

HOW is the subtraction of integers related to the addition of integers?

Use the exercises below to help answer the Inquiry Question.
Write the correct word or phrase on the lines provided.

1. Rewrite the question in your own words.

_____

_____

2. What key words do you see in the question?

_____

Use the following equations to answer Exercises 3-5: $-4 - 7 = -11$ **and** $-4 + (-7) = -11$.

3. What operation is used in the first equation? _____

4. What operation is used in the second equation? _____

5. Is the difference from the first equation equal to the sum from the second equation?

_____

6. Subtracting a negative integer is like adding a _____ .

7. Subtracting a positive integer is like adding a _____ .

8. Write an addition equation that is equivalent to the subtraction equation

$9 - (-8) = 17$. _____

9. Write a subtraction equation that is equivalent to the addition equation

$-6 + (-5) = -11$. _____

HOW is the subtraction of integers related to the addition of integers?

_____

_____

Lesson 3 Notetaking

Subtract Integers

Use the flow chart to review the process for subtracting integers.

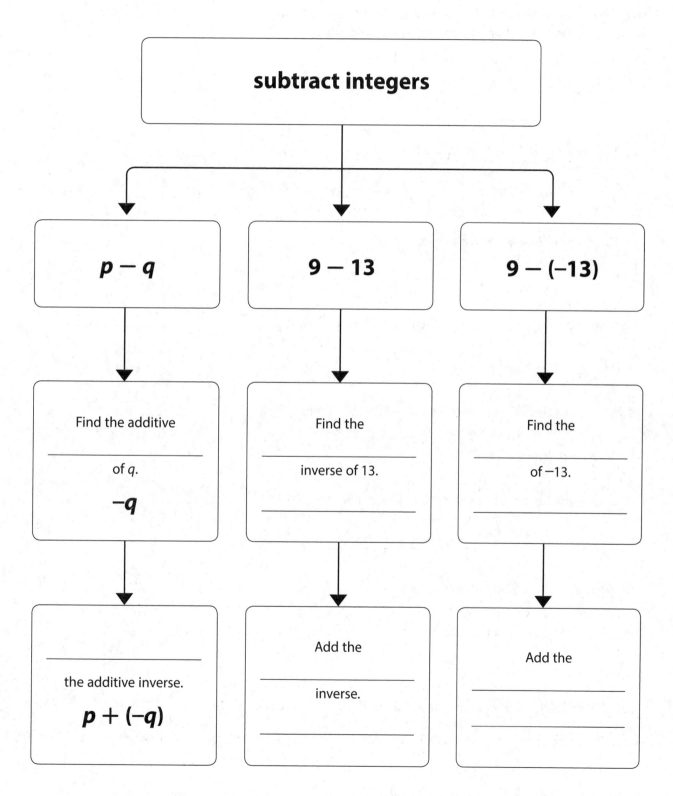

subtract integers

$p - q$	$9 - 13$	$9 - (-13)$

Find the additive _____ of q.

$-q$

Find the _____ inverse of 13. _____

Find the _____ of -13. _____

_____ the additive inverse.

$p + (-q)$

Add the _____ inverse. _____

Add the _____ _____

Inquiry Lab Guided Writing

Distance on a Number Line

HOW is the distance between two rational numbers related to their difference?

Use the exercises below to help answer the Inquiry Question.
Write the correct word or phrase on the lines provided.

1. Rewrite the question in your own words.

_____

_____

2. What key words do you see in the question?

_____

_____

Use the number line below to answer Exercises 3-8.

3. What is the distance between −3 and 1 on the number line? _____

4. Find the difference: −3 − 1 = _____

5. What is the distance between −4 and −2 on the number line? _____

6. Find the difference: −4 − (−2) = _____

7. What is the distance between 5 and 6 on the number line? _____

8. Find the difference: 5 − 6 = _____

9. Are the absolute value of each difference and each distance the same? _____

10. Are all of the distances written as positive or negative numbers? _____

HOW is the distance between two rational numbers related to their difference?

_____

_____

Problem-Solving Investigation
Look for a Pattern

Case 3 Nature

A sunflower usually has two different spirals of seeds, one with 34 seeds and the other with 55 seeds.

The numbers 34 and 55 are part of the Fibonacci sequence.

1, 1, 2, 3, 5, 8, 13, 21, 34, 55, ...

Find the pattern in the Fibonacci sequence and **identify the next two terms**.

- Understand:

- Plan:

- Solve:

- Check:

Case 4 Financial Literacy

Peter is saving money to buy an MP3 player.

After one month, he has **$50.**
After 2 months, he has **$85.**
After 3 months, he has **$120.**
After 4 months, he has **$155.**

At this rate, **how long will it take** Peter **to save** enough money to buy an MP3 player that costs **$295**?

- Understand:

- Plan:

- Solve:

- Check:

Inquiry Lab Guided Writing

Multiply Integers

WHEN is the product of two integers a positive number?
WHEN is the product a negative number?

Use the exercises below to help answer the Inquiry Question.
Write the correct word or phrase on the lines provided.

1. Rewrite the question in your own words.

_____

_____

2. What key words do you see in the question?

_____

3. The answer to a multiplication problem is a _____ .

Use the equations below to answer Exercises 4-7.

a. $-3 \times (-5) = 15$ **b.** $6 \times (-2) = -12$ **c.** $-1 \times 9 = -9$ **d.** $4 \times 7 = 28$

4. Which equations have a product that is positive? _____

5. In which equations do both factors have the same sign (positive or negative)?

_____

6. Which equations have a product that is negative? _____

7. In which equations do the factors have different signs (positive or negative)?

_____

WHEN is the product of two integers a positive number? WHEN is the product a negative number?

_____

_____

_____

Lesson 4 Notetaking

Multiply Integers

Use Cornell notes to better understand the lesson's concepts. Complete each sentence by filling in the blanks with the correct word or phrase.

Questions	Notes
1. What sign is the product of integers with different signs?	The product of two integers with _____ signs is _____ .
2. What sign is the product of integers with the same sign?	The product of two integers with the _____ signs is _____ .

Summary

When is the product of two or more integers a positive number?

_____

_____

_____

_____

_____

_____

_____

_____

Inquiry Lab Guided Writing

Use Properties to Multiply

HOW can properties be used to prove rules for multiplying integers?

Use the exercises below to help answer the Inquiry Question.
Write the correct word or phrase on the lines provided.

1. Rewrite the question in your own words.

_____

_____

2. What key words do you see in the question?

_____

3. Rules for multiplying integers are called multiplication _____ .

4. The Multiplicative Property of Zero states that the product of any number

and zero is _____ .

5. The Multiplicative Identity Property states that the product of any number and

_____ is the number.

6. The Additive Inverse Property states that the sum of an integer and

its _____ is zero.

7. The equation $a(b + c) = ab + ac$ is an example of the _____ .

8. Write some synonyms for *prove*. _____

9. To prove a math statement means to show that it is true for all values.
Which multiplication property proves the statement $0 = 4[3 + (-3)]$.

_____

HOW can properties be used to prove rules for multiplying integers?

_____

_____

_____

Lesson 5 Notetaking

Divide Integers

Use the flow chart to review the process for dividing integers.

```
┌─────────────────────────────────────────┐
│            divide integers              │
└─────────────────────────────────────────┘
```

$p \div q$	$-9 \div 3$	$-9 \div (-3)$
Determine if the signs of p and q are the **same** or **different**.	Are the signs the same or different? _____	Are the signs the same or different? _____

same: positive quotient	**different:** negative quotient	Find the quotient. _____	Find the quotient. _____

Inquiry Lab Guided Writing

Rational Numbers on the Number Line

HOW can you graph negative fractions on the number line?

Use the exercises below to help answer the Inquiry Question.
Write the correct word or phrase on the lines provided.

1. Rewrite the question in your own words.

_____

_____

2. What key words do you see in the question?

_____

3. To _____ a fraction, you find and mark its spot
on a number line.

4. On a number line, negative fractions are to the _____

of zero. Positive fractions are to the _____ of zero.

Use the number line below to answer Exercises 5-8.

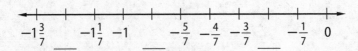

$$-1\frac{3}{7} \quad \underline{\quad} \quad -1\frac{1}{7} \quad -1 \quad \underline{\quad} \quad -\frac{5}{7} \quad -\frac{4}{7} \quad -\frac{3}{7} \quad \underline{\quad} \quad -\frac{1}{7} \quad 0$$

5. Write the missing fractions on the number line.

6. On the number line, what fraction is between $-\frac{3}{7}$ and $-\frac{5}{7}$? _____
Graph the fraction.

7. Write three fractions from the number line that are greater than $-\frac{5}{7}$. _____

8. Write three fractions from the number line that are less than $-\frac{5}{7}$. _____

HOW can you graph negative fractions on the number line?

_____

_____

_____

Lesson 1 Vocabulary

Terminating and Repeating Decimals

Use the vocabulary squares to write a definition, a sentence, and an example for each vocabulary word.

	Definition
repeating decimal	
Example	**Sentence**

	Definition
bar notation	
Example	**Sentence**

	Definition
terminating decimal	
Example	**Sentence**

Lesson 2 Vocabulary

Compare and Order Rational Numbers

Use the three column chart to organize the vocabulary in this lesson.
Write the word in Spanish. Then write the definition of each word.

English	Spanish	Definition
rational number		
integer		
common denominator		
least common denominator (lcd)		

Inquiry Lab Guided Writing

Add and Subtract on the Number Line

HOW can you use a number line to add and subtract like fractions?

Use the exercises below to help answer the Inquiry Question.
Write the correct word or phrase on the lines provided.

1. Rewrite the question in your own words.

_____

_____

2. What key words do you see in the question?

_____

3. The top number of a fraction is the _____ .

4. The bottom number of a fraction is the _____ .

5. What are fractions that have the same denominator called? _____

Use the number line below to answer Exercises 6 and 7.

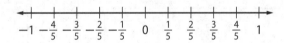

6. To find the sum of $-\frac{3}{5} + \frac{4}{5}$, start at the fraction _____ .

Move 4 units to the _____ to show adding $\frac{4}{5}$.

The sum is _____ .

7. To find the difference for $\frac{1}{5} - \frac{3}{5}$, start at the fraction _____ .

Move 3 units to the _____ to show taking away $\frac{3}{5}$.

The difference is _____ .

HOW can you use a number line to add and subtract like fractions?

_____

_____

_____

Lesson 3 Vocabulary

Add and Subtract Like Fractions

Use the definition map to list qualities about the vocabulary word or phrase.

Vocabulary

like fractions

Description

Characteristics

Examples

Lesson 4 Vocabulary

Add and Subtract Unlike Fractions

Use the concept web to write pairs of unlike fractions.

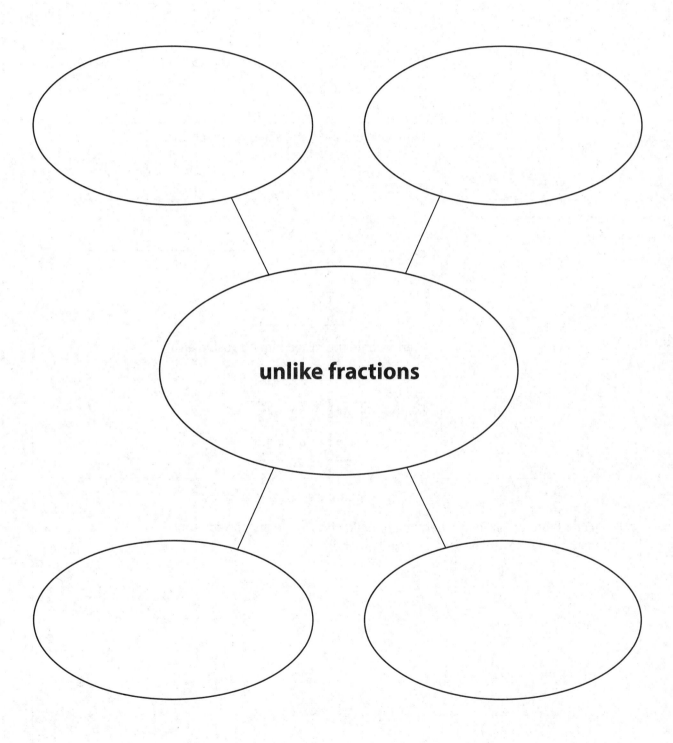

Lesson 5 Notetaking

Add and Subtract Mixed Numbers

Use Cornell notes to better understand the lesson's concepts. Complete each sentence by filling in the blanks with the correct word or phrase.

Questions	Notes
1. How do I add mixed numbers?	First, I _____ the fractions. If necessary, I rename then using the _____ _____ . Then I _____ the whole numbers and simplify if necessary.
2. How do I rename a mixed number?	I can rename the fraction using the _____ . For example: $3\frac{1}{3} \rightarrow 3\frac{2}{6}$ $+2\frac{3}{6} \rightarrow +2\frac{3}{6}$ I can rename the mixed number, using an _____ fraction. For example: $3\frac{2}{6} \rightarrow 2\frac{8}{6}$ $-2\frac{3}{6} \rightarrow -2\frac{3}{6}$

Summary
How can I subtract mixed numbers when the fraction in the first mixed number is less than the fraction in the second mixed number?   _____   _____   _____   _____

Problem-Solving Investigation
Draw a Diagram

Case 3 Fractions

Marta ate a **quarter of** a whole pie.

Edwin ate $\frac{1}{4}$ **of what was left**.

Cristina then ate $\frac{1}{3}$ **of what was left.**

What fraction of the pie **remains**?

- Understand:

- Plan:

- Solve:

- Check:

Case 4 Games

Eight members of a chess club are having a tournament.

In the first round, **every player will play** a chess game **against every other player.**

How many games will be in the first round of the tournament?

- Understand:

- Plan:

- Solve:

- Check:

Lesson 6 Review Vocabulary

Multiply Fractions

Use the flow chart to review the process for multiplying mixed numbers.

multiply mixed numbers
$$2\frac{1}{3} \times \frac{1}{2}$$

Rename the mixed number as an _____ fraction.

$$2\frac{1}{3} = \underline{\qquad}.$$

Multiply the _____.

$$\underline{} \times \frac{1}{2} = \frac{\underline{} \times 1}{\underline{} \times 2} \text{ or } \underline{}$$

Simplify.

$$\underline{\qquad}.$$

Use **mental math** to rename the _____ number.

$$2\frac{1}{3} = \underline{} + \underline{}.$$

Use the _____ to multiply.

$$\left(\underline{} + \underline{}\right) \times \frac{1}{2} = \left(\underline{} \times \frac{1}{2}\right) + \left(\underline{} \times \frac{1}{2}\right) \text{ or } \underline{} + \underline{}$$

Rewrite the sum as a _____ number.

$$\underline{\qquad}.$$

Lesson 7 Notetaking

Convert Between Systems

Use the concept web to show common customary and metric relationships.

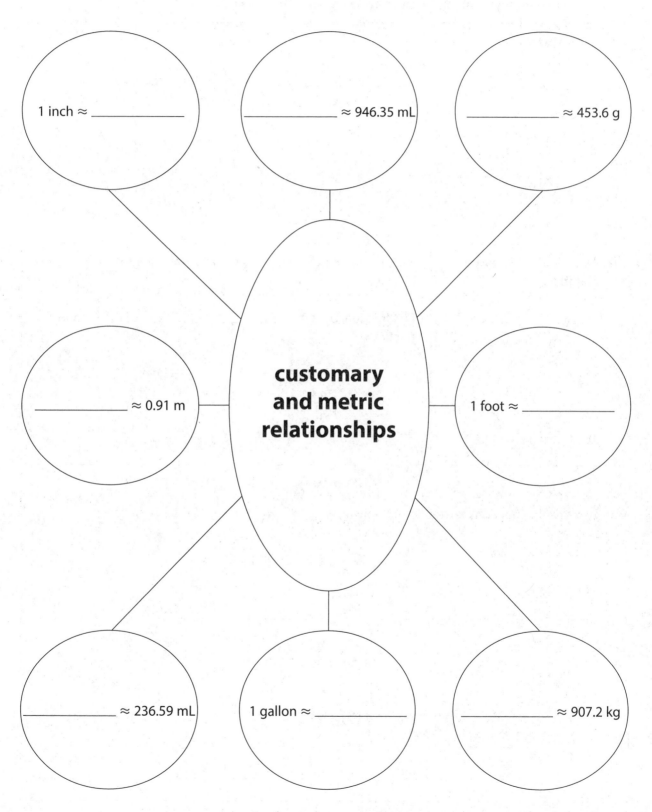

1 inch ≈ _____

_____ ≈ 946.35 mL

_____ ≈ 453.6 g

_____ ≈ 0.91 m

customary and metric relationships

1 foot ≈ _____

_____ ≈ 236.59 mL

1 gallon ≈ _____

_____ ≈ 907.2 kg

Lesson 8 Notetaking

Divide Fractions

Use Cornell notes to better understand the lesson's concepts. Complete each sentence by filling in the blanks with the correct word or phrase.

Questions	Notes
1. How do I divide fractions?	_____ by its multiplicative inverse, or _____ .
2. How do I divide mixed numbers?	First, I rename the mixed number as a fraction greater than one, or an _____ fraction. Then I multiply the first fraction, by the _____ of the second fraction.

Summary

How is dividing fractions related to multiplying?

_____

_____

_____

_____

_____

_____

_____

Lesson 1 Vocabulary

Algebraic Expressions

Use the three column chart to organize the vocabulary in this lesson. Write the word in Spanish. Then write the definition of each word.

English	Spanish	Definition
variable		
algebraic expression		
algebra		
coefficient		
defining a variable		

Lesson 2 Vocabulary
Sequences

Use the definition map to list qualities about the vocabulary word or phrase.

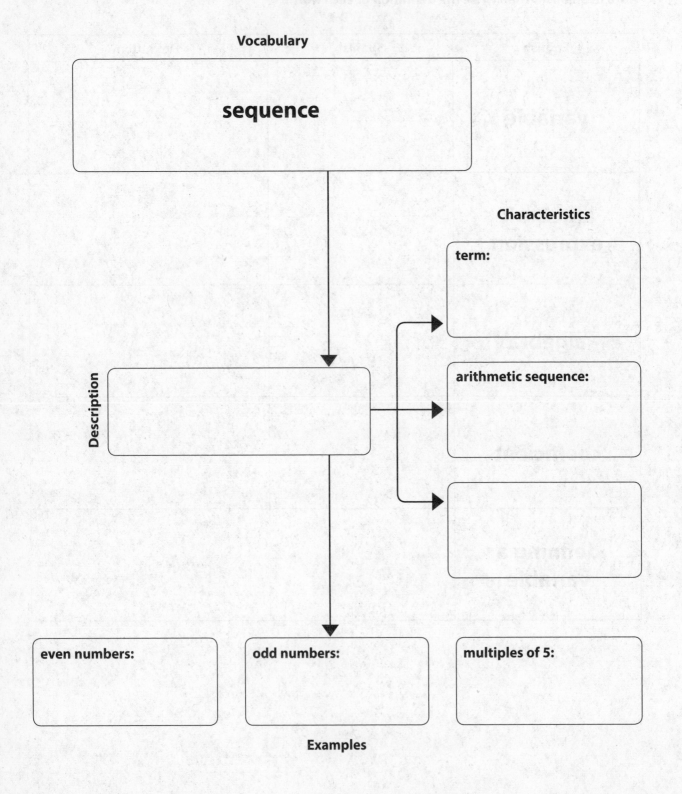

Vocabulary

sequence

Description

Characteristics

term:

arithmetic sequence:

even numbers:

odd numbers:

multiples of 5:

Examples

Inquiry Lab Guided Writing

Sequences

HOW can geometric figures be used to model numerical patterns?

Use the exercises below to help answer the Inquiry Question.
Write the correct word or phrase on the lines provided.

1. Rewrite the question in your own words.

   _____

   _____

2. What key words do you see in the question?

   _____

3. An ordered list of numbers is a _____ .

4. A sequence of numbers that follows a rule is a _____ .

5. Name geometric figures that are polygons.

   _____

Use the pattern below to answer Exercises 6-8.

  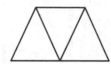

6. Count the number of lines shown in each figure. Write the numerical pattern.

   _____

7. Draw the next figure in the pattern.

8. The geometric figures help you see that the number pattern grows by how many each time?

   _____

HOW can geometric figures be used to model numerical patterns?

_____

_____

_____

Lesson 3 Vocabulary
Properties of Operations

Use the three column chart to organize the vocabulary in this lesson. Write the word in Spanish. Then write the definition of each word.

English	Spanish	Definition
Commutative Property		
Associative Property		
properties		
Additive Identity Property		
Multiplicative Identity Property		
Multiplicative Property of Zero		
counterexample		

Lesson 4 Vocabulary

The Distributive Property

Use the word cards to define each vocabulary word or phrase and give an example.

Word Cards

Distributive Property	propiedad distributiva
Definition	**Definición**
_____	_____
_____	_____
_____	_____
Example Sentence	
_____	
_____	

- -

Word Cards

equivalent expressions	expresiones equivalentes
Definition	**Definición**
_____	_____
_____	_____
_____	_____
Example Sentence	
_____	
_____	

Problem-Solving Investigation
Make a Table

Case 3 Carnivals

For a carnival game, containers are arranged in a **triangular display**.

The **top row has 1** container.
The **second row has 2** containers.
The **third row has 3** containers.

The pattern continues until **the bottom row,** which **has 10** containers.

A contestant **knocks down 29** containers on the first throw.

How many containers **remain**?

- Understand:

- Plan:

- Solve:

- Check:

Case 4 Budget

Tamara **earns $2,050 each month**.

She **spends 65%** of the amount she earns.

The **rest of the money** is **equally divided** and **deposited into two** separate **accounts**.

How many months until Tamara has deposited more than $2,500 in **one** of her accounts?

- Understand:

- Plan:

- Solve:

- Check:

Lesson 5 Vocabulary

Simplify Algebraic Expressions

Use the vocabulary squares to write a definition, a sentence, and an example for each vocabulary word.

term	**Definition**
Example	**Sentence**

like terms	**Definition**
Example	**Sentence**

constant	**Definition**
Example	**Sentence**

Lesson 6 Vocabulary

Add Linear Expressions

Use the definition map to list qualities about the vocabulary word or phrase.

Vocabulary

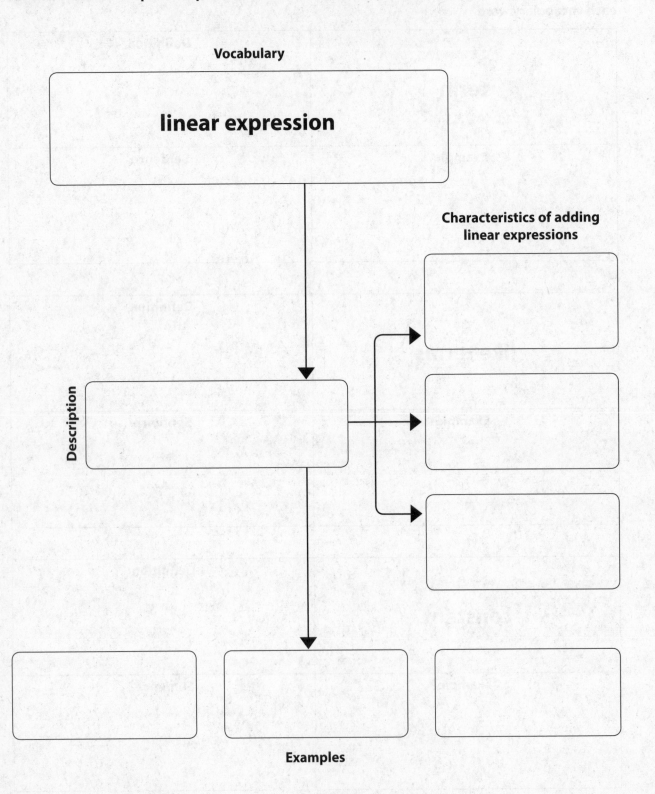

linear expression

**Characteristics of adding
linear expressions**

Description

Examples

Lesson 7 Notetaking

Subtract Linear Expressions

Use Cornell notes to better understand the lesson's concepts. Complete each sentence by filling in the blanks with the correct word or phrase.

Questions	Notes
1. How do I subtract linear expressions?	I subtract _____ terms. I use _____ pairs if needed.
2. What is the additive inverse of a linear expression?	The additive inverse of a linear expression is an expression with terms that are _____ . The sum of a linear expression and its additive inverse is _____ .

Summary
How can I use the additive inverse to help you subtract linear expressions?   _____   _____   _____   _____   _____   _____   _____

Inquiry Lab Guided Writing

Factor Linear Expressions

HOW do models help you factor linear expressions?

Use the exercises below to help answer the Inquiry Question.
Write the correct word or phrase on the lines provided.

1. Rewrite the question in your own words.

_____

_____

2. What key words do you see in the question?

_____

3. Factoring is finding the _____ of a number or expression.

For Examples 4 and 5, write the number of *x*-tiles and 1-tiles or −1-tiles needed to model each expression.

4. $7x - 5$: _____ *x*-tiles   _____ −1-tiles

5. $4x + 3$: _____ *x*-tiles   _____ 1-tiles

Use the algebra tiles below to answer Exercises 6-10.

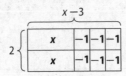

6. Into what shape are the algebra tiles arranged? _____

7. What expression is represented by the whole group of tiles? _____

8. How many rows are there? _____

9. What expression represents each row? _____

10. What is the factored expression? _____

HOW do models help you factor linear expressions?

_____

_____

Lesson 8 Vocabulary
Factor Linear Expressions

Use the vocabulary squares to write a definition, a sentence, and an example for each vocabulary word.

monomial	Definition
Example	**Sentence**

factor	Definition
Example	**Sentence**

factored form	Definition
Example	**Sentence**

Inquiry Lab Guided Writing

Solve One-Step Addition and Subtraction Equations

HOW can bar diagrams or algebra tiles help you solve an equation?

Use the exercises below to help answer the Inquiry Question.
Write the correct word or phrase on the lines provided.

1. Rewrite the question in your own words.

_____

_____

2. What key words do you see in the question?

_____

Use the bar diagram below to answer Exercises 3-7.

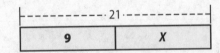

3. What is the total amount shown on the bar diagram? _____

4. In an _____ equation, the total amount is the sum.

5. Write an addition equation that the bar diagram represents.

_____

6. What operation is the inverse of addition? _____

7. Write a subtraction equation that the bar diagram represents.

_____

8. Algebra tiles are another way to _____ equations.

You can add tiles to show _____ and take away tiles

to show _____.

HOW can bar diagrams or algebra tiles help you solve an equation?

_____

_____

Lesson 1 Vocabulary

Solve One-Step Addition and Subtraction Equations

Use the three column chart to organize the vocabulary in this lesson. Write the word in Spanish. Then write the definition of each word.

English	Spanish	Definition
equation		
solution		
equivalent equations		
Subtraction Property of Equality		
Addition Property of Equality		

Inquiry Lab Guided Writing

Multiplication Equations with Bar Diagrams

HOW do you know which operation to use when solving an equation?

Use the exercises below to help answer the Inquiry Question.
Write the correct word or phrase on the lines provided.

1. Rewrite the question in your own words.

_____

_____

2. What key words do you see in the question?

_____

Use the bar diagram below to answer Exercises 3-7.

3. What is the total amount shown on the bar diagram? _____

4. The bar diagram shows that h is divided into how many equal parts? _____

5. Write a division sentence that is represented by the bar diagram.

_____

6. What operation is the inverse of division? _____

7. Write a multiplication sentence represented by the bar diagram.

_____

8. You can use _____ , or opposite, operations to solve equations.

HOW do you know which operation to use when solving an equation?

_____

_____

_____

Lesson 2 Vocabulary

Multiplication and Division Equations

Use the vocabulary squares to write a definition, a sentence, and an example for each vocabulary word.

	Definition
coefficient	
Example	Sentence

	Definition
Division Property of Equality	
Example	Sentence

	Definition
Multiplication Property of Equality	
Example	Sentence

Inquiry Lab Guided Writing

Solve Equations with Rational Coefficients

HOW can you use bar diagrams to solve equations with rational coefficients?

Use the exercises below to help answer the Inquiry Question.
Write the correct word or phrase on the lines provided.

1. Rewrite the question in your own words.

_____

_____

2. What key words do you see in the question?

_____

3. In the term $\frac{3}{4}x$, the fraction $\frac{3}{4}$ is the _____ .

The variable is _____ .

4. To model the fraction $\frac{3}{4}$, into how many equal sections would you divide a bar diagram?

_____

Use the bar diagram below to answer Exercises 5-7.

5. How many equal sections does the number 36 cover? _____

What fraction of the total does 36 represent? _____

6. Complete the division equation that shows the value of

one section: $36 \div$ _____ $= 12$.

7. If one section, or $\frac{1}{4}$, of the total bar diagram is equal to 12, you can

multiply by _____ to find the total value.

8. Write a multiplication sentence that shows the value of the whole bar. _____

HOW can you use bar diagrams to solve equations with rational coefficients?

_____

_____

Lesson 3 Notetaking

Solve Equations with Rational Coefficients

Use Cornell notes to better understand the lesson's concepts. Complete each sentence by filling in the blanks with the correct word or phrase.

Questions	Notes
1. How do I solve an equation with a decimal coefficient?	I can use the _____ Property of Equality and _____ each side by the coefficient.
2. How do I solve an equation with a fractional coefficient?	I can use the _____ Property of Equality and _____ each side by the _____ of the coefficient.

Summary

What is the process for solving a multiplication equation with a rational coefficient?

_____

_____

_____

_____

_____

_____

_____

Inquiry Lab Guided Writing

Solve Two-Step Equations

HOW can a bar diagram or algebra tiles help you solve a real-world problem?

**Use the exercises below to help answer the Inquiry Question.
Write the correct word or phrase on the lines provided.**

1. Rewrite the question in your own words.

_____

_____

2. What key words do you see in the question?

_____

3. Name a type of model that you can divide into parts to represent a whole.

_____

4. What type of model has pieces that you can move, add, and subtract

to represent parts of an equation? _____

5. Which type of model helps you work through a problem step-by-step? _____

6. How are a bar diagram and algebra tiles alike?

_____

_____

7. How are a bar diagram and algebra tiles different?

_____

_____

_____

HOW can a bar diagram or algebra tiles help you solve a real-world problem?

_____

_____

_____

Lesson 4 Vocabulary

Solve Two-Step Equations

Use the definition map to list qualities about the vocabulary word or phrase.

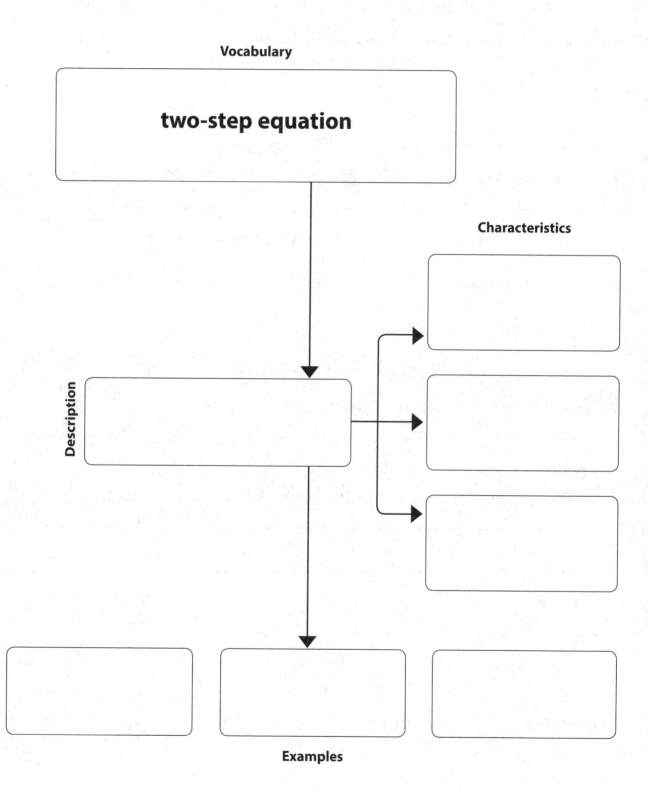

Vocabulary

two-step equation

Description

Characteristics

Examples

Inquiry Lab Guided Writing

More Two-Step Equations

HOW are equations in $p(x + q) = r$ form different from $px + q = r$ equations?

Use the exercises below to help answer the Inquiry Question.
Write the correct word or phrase on the lines provided.

1. Rewrite the question in your own words.

_____

_____

2. What key words do you see in the question?

_____

For Exercises 3-5, complete each sentence to identify the order of operations.

3. First, perform operations inside _____ .

4. Then, multiply and _____ from left to right.

5. Finally, _____ and subtract from left to right.

6. How are the equations in the Inquiry Question different?

_____

For Exercises 7-9, let $p = 2$, $x = 3$, and $q = 4$.

7. What is the value of r in $p(x + q) = r$? Show your work below. _____

_____

8. What is the value of r in $px + q = r$? Show your work below. _____

_____

9. In the first equation, why did you add before multiplying?

_____

HOW are equations in $p(x + q) = r$ form different from $px + q = r$ equations?

_____

_____

_____

Lesson 5 Review Vocabulary

More Two-Step Equations

Use the concept web to write and solve examples of two-step equations. Include examples written in the forms $px + q = r$ and $p(x + q) = r$.

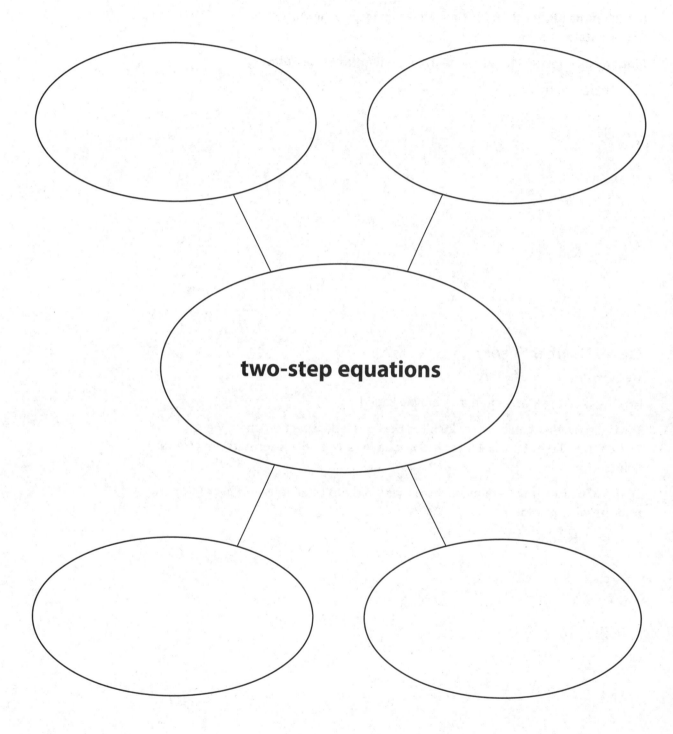

Problem-Solving Investigation
Work Backward

Case 3 Waterfalls

Angel Falls in Venezuela is **3,212 feet high**.

It is **29 yards higher than 2.5 times** the architectural **height of the Empire State Building**.

Find the architectural **height**, in feet, **of the Empire State Building**.

- Understand:

- Plan:

- Solve:

- Check:

Case 4 Number Theory

Travis works at a kite factory.

He checks all the kites before they are packaged.

Travis discovered that for every **28 kites that passed** inspection, there were **7 kites that did not pass: 4 kites did not have tails, and 3 kites had the wrong colors**.

Of the **476 kites** Travis examined, **how many did not have tails** and **how many had the wrong colors?**

- Understand:

- Plan:

- Solve:

- Check:

Inquiry Lab Guided Writing

Solve Inequalities

HOW is an inequality like an equation? How is it different?

Use the exercises below to help answer the Inquiry Question.
Write the correct word or phrase on the lines provided.

1. Rewrite the question in your own words.

_____

_____

2. What key words do you see in the question?

_____

3. The equation $y + 2 = 5$ compares the quantities _____ and

_____ .

4. The $=$ symbol means the quantities $y + 2$ and 5 are _____ .

5. What symbols are used to show two quantities are not equal? _____

6. What does the $\leq$ symbol mean? _____

7. What does the $\geq$ symbol mean? _____

8. What quantities are being compared in the inequality $x + 3 > 7$? _____

9. How would the number sentence change if the value of x was 4?

_____

_____

HOW is an inequality like an equation? How is it different?

_____

_____

_____

Lesson 6 Vocabulary

Solve Inequalities by Addition or Subtraction

Use the vocabulary squares to write a definition, a sentence, and an example for each vocabulary word.

	Definition
inequality	
Example	Sentence

	Definition
Subtraction Property of Inequality	
Example	Sentence

	Definition
Addition Property of Inequality	
Example	Sentence

Lesson 7 Notetaking

Solve Inequalities by Multiplication or Division

Use Cornell notes to better understand the lesson's concepts. Complete each sentence by filling in the blanks with the correct word or phrase.

Questions	Notes
1. What do the Multiplication Property of Inequality and the Division Property of Inequality state about positive numbers?	That an inequality _____ when you _____ or _____ each side of an inequality by a _____ number.
2. What do the Multiplication Property of Inequality and the Division Property of Inequality state about negative numbers?	That an the inequality _____ must be _____ when you _____ or _____ each side of an inequality by a _____ number.

Summary
When should I reverse the inequality symbol when solving an inequality?   _____   _____   _____   _____   _____   _____

Lesson 8 Vocabulary

Solve Two-Step Inequalities

Use the definition map to list qualities about the vocabulary word or phrase.

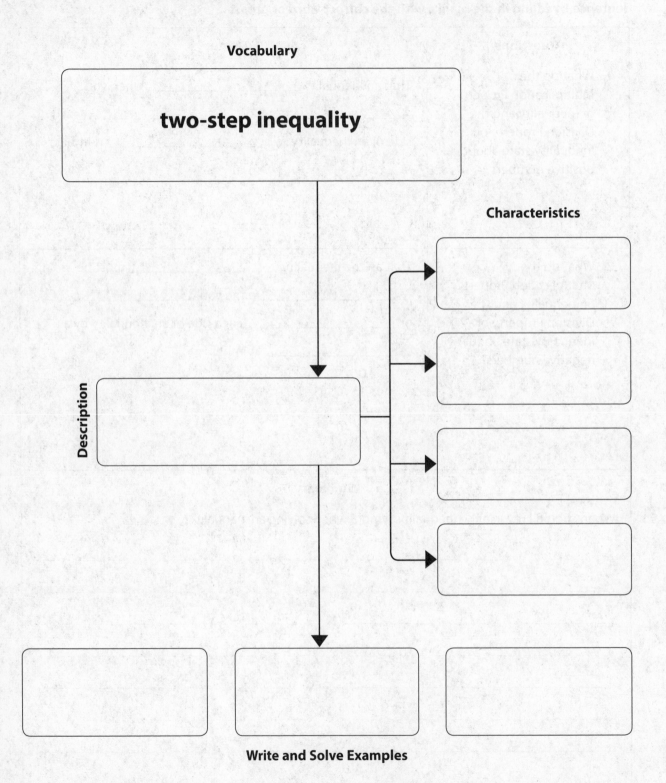

Vocabulary

two-step inequality

Description

Characteristics

Write and Solve Examples

Lesson 1 Vocabulary

Classify Angles

Use the three column chart to write the vocabulary word and definition for each drawing.

What I See	Vocabulary Word	Definition

Lesson 2 Vocabulary
Complementary and Supplementary Angles

Use the word cards to define each vocabulary word or phrase and give an example.

Word Cards

complementary angles	ángulos complementarios
Definition	**Definición**
Example Sentence	

Word Cards

supplementary angles	ángulos suplementarios
Definition	**Definición**
Example Sentence	

Inquiry Lab Guided Writing

Create Triangles

WHAT do you notice about the measures of the sides or the measures of the angles that form triangles?

Use the exercises below to help answer the Inquiry question.
Write the correct word or phrase on the lines provided.

1. Rewrite the question in your own words.

_____

_____

2. What key words do you see in the question?

_____

3. Use the given triangles to fill in the table below.
Find the sum of two side lengths of each triangle.
Then write the length of the third side.

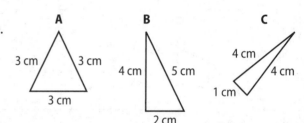

Triangle	Sum of sides 1 and 2	Length of side 3
A		
B		
C		

The angle measures for three triangles are listed. Find each sum.

4. $60° + 30° + 90° =$ _____

5. $85° + 25° + 70° =$ _____

6. $55° + 75° + 50° =$ _____

WHAT do you notice about the measures of the sides or the measures of
the angles that form triangles?

_____

_____

_____

Lesson 3 Vocabulary
Triangles

Use the three column chart to write the vocabulary word and definition for each drawing.

What I See	Vocabulary Word	Definition
Name by **angles**		
Name by **sides**		
General vocabulary		

Inquiry Lab Guided Writing

Draw Triangles

HOW can you use technology to draw geometric shapes?

Use the exercises below to help answer the Inquiry Question.
Write the correct word or phrase on the lines provided.

1. Rewrite the question in your own words.

_____

_____

2. What key words do you see in the question?

_____

3. What is a polygon?

_____

_____

4. What polygon has three sides and three angles? _____

5. The rays that form an angle meet at the _____ .

The plural of *vertex* is _____ .

6. Geometry software can help you draw polygons.

The Straightedge tool is used for drawing _____ .

7. The _____ tool is used to check the measures of sides and angles.

8. You can move vertices to change the sizes of _____ and

_____ .

HOW can you use technology to draw geometric shapes?

_____

_____

_____

Problem-Solving Investigation
Make a Model

Case 3 Tables

Members of Student Council are setting up **tables end-to-end** for an awards banquet.

Each table will seat **one person on each side.**

How many **square tables** will they need to put together for **32 people**?

Explain your method.

- Understand:

- Plan:

- Solve:

- Check:

Case 4 Tile

The diagram shows the design of a tile border around a rectangular swimming pool that measures **10.5 meters by 6 meters**.

Each tile is a **square** measuring **1.5 meters on each side**.

Explain a method you could use to **find the area** of just **the tile border**. Then **solve**.

- Understand:

- Plan:

- Solve:

- Check:

Inquiry Lab Guided Writing

Investigate Online Maps and Scale Drawings

HOW is the zoom feature of an online map like the scale of a drawing?

Use the exercises below to help answer the Inquiry Question.
Write the correct word or phrase on the lines provided.

1. Rewrite the question in your own words.

_____

_____

2. What key words do you see in the question?

_____

3. A _____ is used when drawing objects that are too large or small to be drawn at actual size.

Use the maps below to answer Exercises 4-6.

A

B

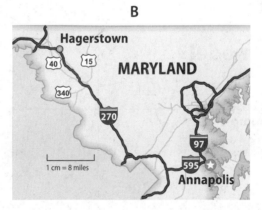

4. Which map shows what it is like to use the zoom in feature? _____

5. What are the scales for the maps?

   A: _____   **B:** _____

6. As you zoom in, a map looks _____ . As you zoom out, a map looks

   _____ .

HOW is the zoom feature of an online map like the scale of a drawing?

_____

_____

_____

Lesson 4 Vocabulary

Scale Drawings

Use the vocabulary squares to write a definition, a sentence, and an example for each vocabulary word.

	Definition
scale drawing	
Example	**Sentence**

	Definition
scale model	
Example	**Sentence**

	Definition
scale factor	
Example	**Sentence**

Inquiry Lab Guided Writing

Scale Drawings

WHAT happens to the size of a scale drawing when it is reproduced using a different scale?

Use the exercises below to help answer the Inquiry Question.
Write the correct word or phrase on the lines provided.

1. Rewrite the question in your own words.

_____

_____

2. What key words do you see in the question?

_____

Use the scale drawings below to answer Exercises 3-5.

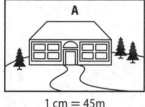

1 cm = 45m

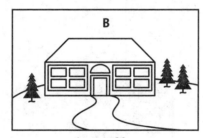

1 cm = 30m

3. What does a single unit represent in Drawing A? _____

4. What does a single unit represent in Drawing B? _____

5. Which scale represents the greater length? _____

6. When the scale length is greater, the scale drawing is _____ .

7. When the scale length is lesser, the scale drawing is _____ .

WHAT happens to the size of a scale drawing when it is reproduced using
a different scale?

_____

_____

Lesson 5 Review Vocabulary

Draw Three-Dimensional Figures

Use the concept web to identify the sides of the three-dimensional figure shown. Draw the views of each side using squares.

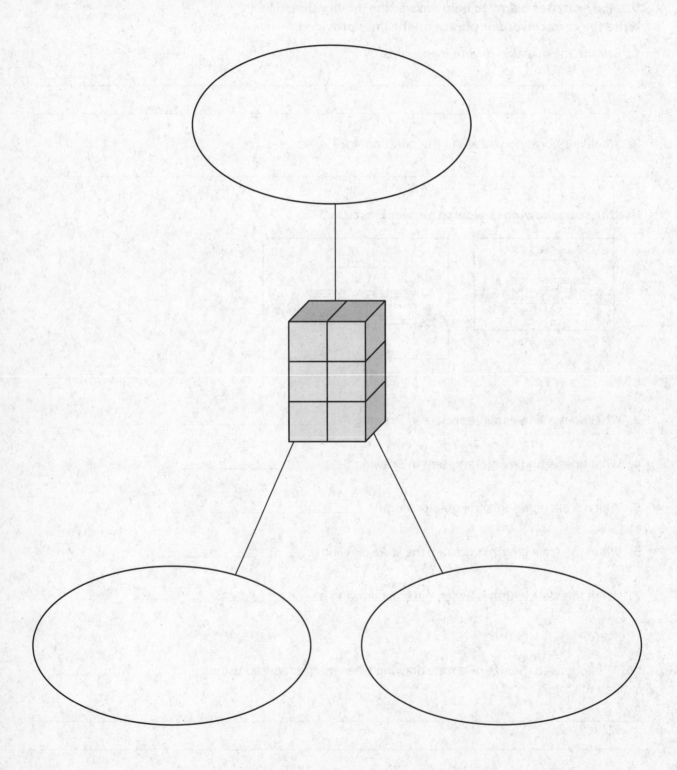

Lesson 6 Vocabulary

Cross Sections

Use the three column chart to organize the vocabulary in this lesson.
Write the word in Spanish. Then write the definition of each word.

English	Spanish	Definition
prism		
base		
pyramid		
plane		
coplanar		
parallel lines		
vertex		
diagonal		
cylinder		
cone		

Inquiry Lab Guided Writing
Circumference

HOW is the circumference of a circle related to its diameter?

Use the exercises below to help answer the Inquiry Question. Write the correct word or phrase on the lines provided.

1. Rewrite the question in your own words.

_____

_____

2. What key words do you see in the question?

_____

3. _____ is the distance around a circle.

4. _____ is the distance across a circle through its center.

5. What does the ≈ symbol mean? _____

6. The circumference of a circle is 47.1 cm. The diameter of the circle is 15 cm.

About how many times greater is the circumference? _____

7. Complete the table by estimating the missing circumference or diameter.

Diameter	Circumference
8 cm	cm
cm	33 cm
cm	15 cm
12 cm	cm

HOW is the circumference of a circle related to its diameter?

_____

_____

_____

Lesson 1 Vocabulary

Circumference

Use the three column chart to write the vocabulary word and definition for each drawing.

What I See	Vocabulary Word	Definition
◯		
⊙		
◯ (dashed)		
⊖		
◯ (radius)		
π		

Inquiry Lab Guided Writing

Area of Circles

HOW are the circumference and area of a circle related?

Use the exercises below to help answer the Inquiry Question. Write the correct word or phrase on the lines provided.

1. Rewrite the question in your own words.

_____

_____

2. What key words do you see in the question?

_____

3. The distance around a circle is its _____ .

4. _____ is the distance from the center to any point on a circle.

5. What is the approximate value of π? _____

6. $A = \pi r^2$ is the formula for finding the _____ of a circle.

Use the figures below to answer Exercises 7-9.

$C = 18.84$ cm

$r = 3$ cm

height $= 3$ cm

base $= 9.42$ cm

7. The sections of a circle can be rearranged to form a _____ .

8. What is the circumference of the circle? _____

9. The _____ of the parallelogram is $\frac{1}{2}$ the circumference of the circle.

HOW are the circumference and area of a circle related?

_____

_____

Lesson 2 Vocabulary

Area of Circles

Use the definition map to list qualities about the vocabulary word or phrase.

Vocabulary

semicircle

Description

Characteristics

Draw examples of semicircles.

Lesson 3 Notetaking

Area of Composite Figures

Use Cornell notes to better understand the lesson's concepts. Complete each sentence by filling in the blanks with the correct word or phrase.

Questions	Notes
1. How do I find the area of a composite figure?	Since a composite figure is made up of two or more _____ , decompose the composite figure. Decompose it into shapes with known _____ formulas. Then find the _____ of these _____ .
2. How do I find the area of a shaded region?	Use shapes with _____ that are known. For example, find the area of a region larger than the shaded region and _____ the non-shaded regions.

Summary
How do measurements help you describe real-world objects?
_____
_____
_____
_____
_____
_____
_____

Lesson 4 Vocabulary

Volume of Prisms

Use the flow chart to review the processes for finding the volume of a prism.

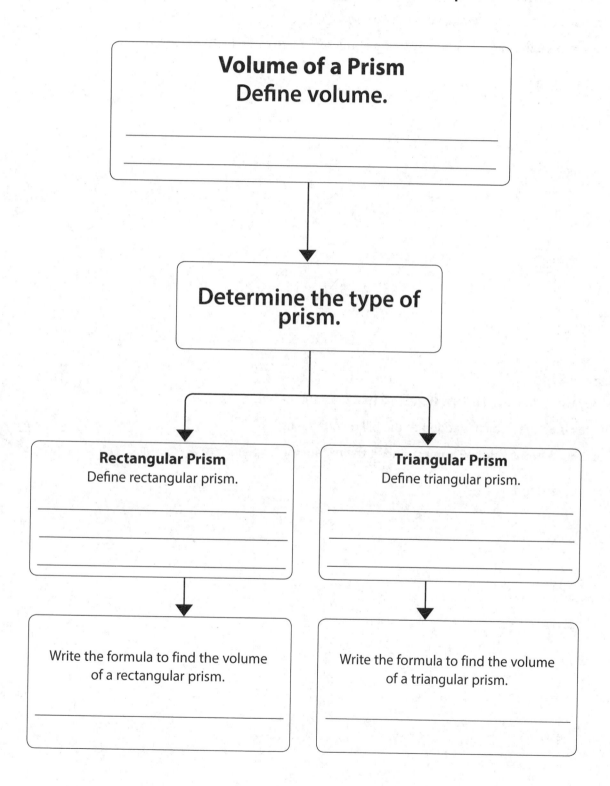

Volume of a Prism
Define volume.

_____

_____

Determine the type of prism.

Rectangular Prism
Define rectangular prism.

_____

_____

_____

Triangular Prism
Define triangular prism.

_____

_____

_____

Write the formula to find the volume of a rectangular prism.

_____

Write the formula to find the volume of a triangular prism.

_____

Problem-Solving Investigation
Solve a Simpler Problem

Case 3 Woodworking

Two workers can make **two chairs** in **two days**.

How many chairs can **8 workers** working at the same rate make in **20 days**?

- Understand:

- Plan:

- Solve:

- Check:

Case 4 Tips

Ebony wants to leave an **18% tip** for a **$19.82** restaurant **bill**.

The tax is **6.25%**, which is added to the bill before the tip.

How much money does Ebony spend at the restaurant? Explain.

- Understand:

- Plan:

- Solve:

- Check:

Inquiry Lab Guided Writing

Volume of Pyramids

WHAT is the relationship between the volume of a prism and the volume of a pyramid with the same base area and height?

Use the exercises below to help answer the Inquiry Question. Write the correct word or phrase on the lines provided.

1. Rewrite the question in your own words.

_____

_____

2. What key words do you see in the question?

_____

3. Write the name of the figure shown.

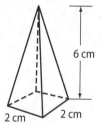

6 cm

2 cm   2 cm

_____

4. Write the name of the figure shown.

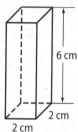

6 cm

2 cm   2 cm

2 cm

_____

5. Are the heights of the figures the same? _____

Are the bases of the figures the same? _____

6. The formula for finding the _____ of a rectangular prism

is $V = \ell wh$ or $V = bh$. The volume of the prism is _____ .

7. The volume of the pyramid is 8 cm³. What fraction represents the volume of

the pyramid compared to the volume of the prism? _____

WHAT is the relationship between the volume of a prism and the volume of a pyramid with the same base area and height?

_____

_____

_____

Lesson 5 Vocabulary
Volume of Pyramids

Use the word cards to define each vocabulary word or phrase and give an example.

Word Cards

pyramid	pirámide

Definition

_____

_____

_____

Definición

_____

_____

_____

Example Sentence

_____

_____

Word Cards

lateral face	cara lateral

Definition

_____

_____

Definición

_____

_____

Example Sentence

_____

_____

Inquiry Lab Guided Writing

Nets of Three-Dimensional Figures

HOW can models and nets help you find the surface area of prisms?

Use the exercises below to help answer the Inquiry Question. Write the correct word or phrase on the lines provided.

1. Rewrite the question in your own words.

_____

_____

2. What key words do you see in the question?

_____

3. A _____ is any surface that forms a side or a base of a prism.

4. The _____ of a prism is the sum of the areas of all the faces.

Use the models below to answer Exercises 5-7.

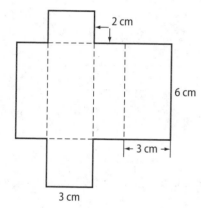
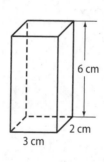

5. What do we call the figure shown on the left? _____

6. What is the name of the figure shown on the right? _____

7. Which figure makes it easier to see and count all the faces of the

rectangular prism? _____

HOW can models and nets help you find the surface area of prisms?

_____

_____

_____

Lesson 6 Vocabulary
Surface Area of Prisms

Use the definition map to list qualities about the vocabulary word or phrase.

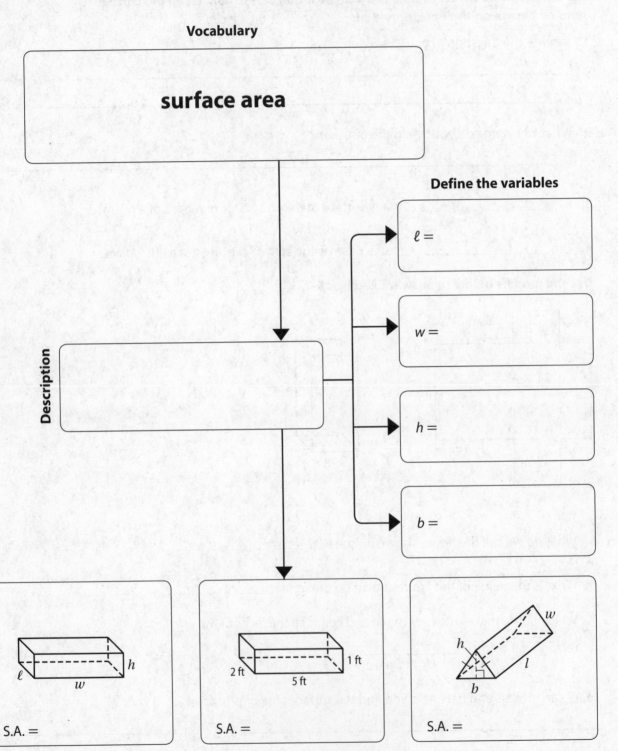

Vocabulary

surface area

Description

Define the variables

$\ell =$

$w =$

$h =$

$b =$

S.A. =

S.A. =

S.A. =

Find the surface area of each figure.

Inquiry Lab Guided Writing

Relate Surface Area and Volume

HOW does the shape of a rectangular prism affect its volume and surface area?

Use the exercises below to help answer the Inquiry Question. Write the correct word or phrase on the lines provided.

1. Rewrite the question in your own words.

   _____

   _____

2. What key words do you see in the question?

   _____

Use the rectangular prisms below to complete the Exercises 3-8.

A   B

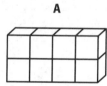

3. How many cubes make up Figure A? _____

   How many cubes make up Figure B? _____

4. Do the figures have the same volume? _____

5. What is the surface area of Figure A? _____

   What is the surface are of Figure B? _____

6. Do the figures have the same surface area? _____

7. Which figure has more faces touching on the interior? _____

8. Which is greater, the surface area of Figure B or the surface area of Figure A?

   _____

HOW does the shape of a rectangular prism affect its volume and surface area?

_____

_____

Lesson 7 Vocabulary
Surface Area of Pyramids

Use the vocabulary squares to write a definition and a sentence for each vocabulary word.

	Definition
lateral surface area	
Write the formula.	**Sentence**

	Definition
slant height	
Draw an arrow showing the slant height.	**Sentence**

	Definition
lateral face	
Draw an arrow showing a lateral face.	**Sentence**

Inquiry Lab Guided Writing

Composite Figures

HOW can you find the volume and surface area of a composite figure?

Use the exercises below to help answer the Inquiry Question. Write the correct word or phrase on the lines provided.

1. Rewrite the question in your own words.

_____

_____

2. What key words do you see in the question?

_____

3. A _____ figure is made up of two or more three-dimensional figures.

Use the figure below to answer Exercises 4-8.

4. What two figures make up the composite figure?

_____

5. What is the formula for finding the volume of a rectangluar prism? _____

6. What is the formula for finding the volume of a pyramid? _____

7. To find the volume of the composite figure, _____ the volumes of the figures that make up the composite figure.

8. Are the faces where the figures touch part of the surface area of the composite figure? _____

HOW can you find the volume and surface area of a composite figure?

_____

_____

_____

Lesson 8 Notetaking

Volume and Surface Area of Composite Figures

Use Cornell notes to better understand the lesson's concepts. Complete each sentence by filling in the blanks with the correct word or phrase.

Questions	Notes
1. How do I find the volume of a composite figure?	Since a composite figure is made up of two or more _____ , decompose the composite figure. Separate it into solids whose _____ formulas are known. Then find the _____ of these _____ .
2. How do I find the surface area of a composite figure?	Find the _____ of the _____ that make up the composite figure.

Summary

How did the lessons in this chapter help you find the surface area and volume of a composite figure?

_____

_____

_____

_____

_____

Lesson 1 Vocabulary
Probability of Simple Events

Use the three column chart to organize the vocabulary in this lesson. Write the word in Spanish. Then write the definition of each word.

English	Spanish	Definition
probability		
outcome		
simple event		
random		
complementary events		

Inquiry Lab Guided Writing

Relative Frequency

HOW is probability related to relative frequency?

Use the exercises below to help answer the Inquiry Question. Write the correct word or phrase on the lines provided.

1. Rewrite the question in your own words.

_____

_____

2. What key words do you see in the question?

_____

3. _____ is the chance that some event will happen.

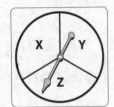

4. Probability is written as a _____ of the number of favorable outcomes to the number of possible outcomes.

5. What is the probability of the given spinner landing on X? _____

Use the experiment results in the table below to answer Exercises 6-8.

Number of Spins	X	Y	Z
10	6	2	2
20	8	8	4
30	11	9	10

6. The _____ frequency of landing on X after 10 spins is $\frac{6}{10}$.

7. What is the relative frequency of landing on X after 20 spins? _____

After 30 spins? _____

8. As the number of spins increases, is the relative frequency closer to

or farther from the probability? _____

HOW is probability related to relative frequency?

_____

_____

_____

Lesson 2 Vocabulary

Theoretical and Experimental Probability

Use the vocabulary squares to write a definition, a sentence, and an example for each vocabulary word.

	Definition
uniform probability model	
Example	**Sentence**

	Definition
theoretical probability	
Example	**Sentence**

	Definition
experimental probability	
Example	**Sentence**

Inquiry Lab Guided Writing

Fair and Unfair Games

HOW can you determine if a game is fair?

Use the exercises below to help answer the Inquiry Question. Write the correct word or phrase on the lines provided.

1. Rewrite the question in your own words.

_____

_____

2. What key words do you see in the question?

_____

3. If you have an advantage in a game, do you have a greater or lesser chance

of winning? _____

A game involves tossing a ball in a basket. The player who makes the most baskets wins.

4. Determine if Player 1 or Player 2 has an advantage in each situation:

a. Player 1 stands 5 feet from the basket. Player 2 stands 8 feet from the basket.

_____

b. The baskets are the same size. Player 1 uses a large ball. Player 2 uses a small ball.

_____

c. Player 1 shoots with open eyes. Player 2 shoots with closed eyes.

_____

5. Do any of the above situations make the game fair? _____

6. Describe the factor in each situation that gives one player an advantage over the other.

a. _____

b. _____

c. _____

HOW can you determine if a game is fair?

_____

_____

_____

Lesson 3 Vocabulary

Probability of Compound Events

Use the flow chart to review the process for creating a tree diagram for a compound event.

Define the phrase **compound event**.

_____

_____

Define the phrase **sample space**.

_____

_____

Define the phrase **tree diagram**.

_____

_____

Compound Event

A sandwich can be made with whole wheat or whole grain bread and three kinds of filling: peanut butter, cheese, or tuna salad.

List the Sample Space for the Compound Event

Use W for whole wheat, G for whole grain, P for peanut butter, C for cheese, and T for tuna salad.

_____

Complete the tree diagram.

| Bread | Filling | Sample Space |

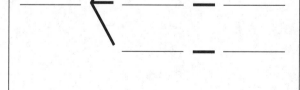

Lesson 4 Vocabulary

Simulations

Use the concept web to define simulation. Then give examples of different simulations from the book.

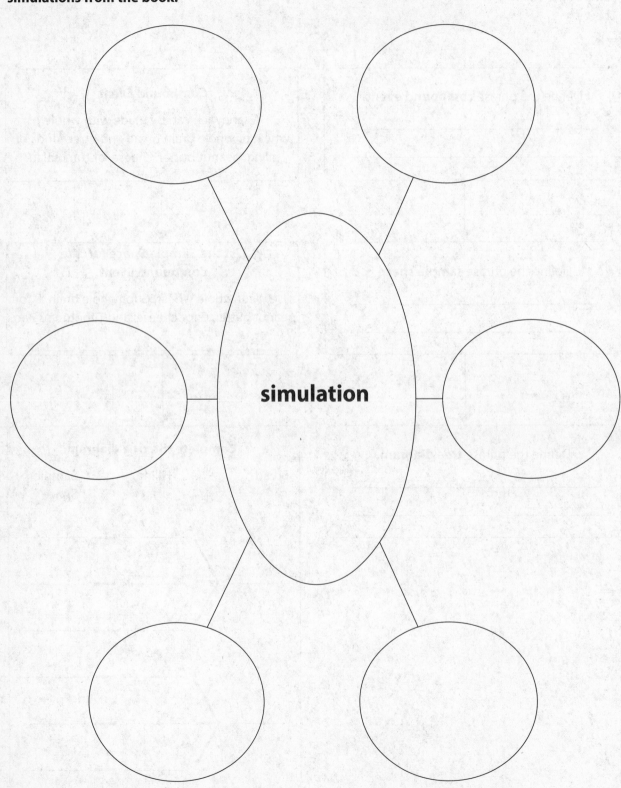

Inquiry Lab Guided Writing

Simulate Compound Events

HOW do simulations help you understand the probability of events happening?

Use the exercises below to help answer the Inquiry Question. Write the correct word or phrase on the lines provided.

1. Rewrite the question in your own words.

_____

_____

2. What key words do you see in the question?

_____

3. A _____ is an experiment that is designed to model a real situation.

4. Name one reason you would use a simulation?

_____

_____

5. How are spinners useful in simulations?

_____

_____

6. Theoretical _____ is what should happen when conducting a probability experiment.

7. _____ probability is what does happen when conducting a probability experiment.

8. As the number of trials increases, does the experimental probability

become closer to or farther from the theoretical probability? _____

HOW do simulations help you understand the probability of events happening?

_____

_____

_____

Problem-Solving Investigation

Act it Out

Case 3 Chess

A chess tournament will be held and **32 students** will participate.

If a **player loses one match**, he or she will be **eliminated.**

How many total games will be played in the tournament?

- Understand:

- Plan:

- Solve:

- Check:

Case 4 Running

Six runners are entered in a race.

Assume there are **no ties**.

In **how many ways can first and second places be awarded**?

- Understand:

- Plan:

- Solve:

- Check:

Lesson 5 Vocabulary
Fundamental Counting Principle

Use the definition map to list qualities about the vocabulary word or phrase.

Vocabulary

Fundamental Counting Principle

Description

**Characteristics:
What it is**

3 choices of bread 8 choices of filling	flipping a dime, quarter, and nickel

a spinner has
8 outcomes
a cube has 6 outcomes

Find the number of outcomes in each situation.

Lesson 6 Notetaking
Permutations

Use Cornell notes to better understand the lesson's concepts. Complete each sentence by filling in the blanks with the correct word or phrase.

Questions	Notes
1. What is a permutation?	A permutation is an _____ , or _____ , of objects in which _____ is important.
2. How do I use permutations to find probability?	I can use _____ to find the number of _____ . This will be the _____ in the probability ratio.

Summary

How can I find the number of permutations of a set of objects?

_____

_____

_____

_____

_____

_____

_____

Inquiry Lab Guided Writing

Independent and Dependent Events

HOW can one event impact a second event in a probability experiment?

Use the exercises below to help answer the Inquiry Question. Write the correct word or phrase on the lines provided.

1. Rewrite the question in your own words.

_____

_____

2. What key words do you see in the question?

_____

There are 2 blue pens, 2 black pens, and 2 red pens in a cup.

3. Tomás takes a pen without looking. What is the probability that he will take

a red pen? _____

4. Tomás puts the first pen back in the cup and takes another pen without

looking. What is the probability that he will choose a red pen this time? _____

5. Why didn't the probability change? _____

6. All pens are back in the cup. Janice takes a pen without looking. What is the

probability that she will take a blue pen? _____

7. Janice takes a blue pen and keeps it. She chooses another pen from the cup.

What is the probability that she will choose a blue pen again? _____

8. Why did the probability change? _____

9. Who demonstrated how a first event can affect what happens in a

second event: Tomás or Janice? _____

HOW can one event impact a second event in a probability experiment?

_____

_____

Lesson 7 Vocabulary

Independent and Dependent Events

Use the word cards to define each vocabulary word or phrase and give an example.

Word Cards

independent events

Definition

_____

_____

_____

Example Sentence

_____

_____

eventos independientes

Definición

_____

_____

_____

Word Cards

dependent events

Definition

_____

_____

_____

Example Sentence

_____

_____

eventos dependientes

Definición

_____

_____

_____

Lesson 1 Vocabulary

Make Predictions

Use the vocabulary squares to write a definition, a sentence, and an example for each vocabulary word.

statistics	Definition
Example	**Sentence**

survey	Definition
Example	**Sentence**

population	Definition
Example	**Sentence**

Lesson 2 Vocabulary

Unbiased and Biased Samples

Use the three column chart to organize the vocabulary in this lesson. Write the word in Spanish. Then write the definition of each word.

English	Spanish	Definition
unbiased sample		
simple random sample		
systematic random sample		
biased sample		
convenience sample		
voluntary response sample		

Inquiry Lab Guided Writing

Multiple Samples of Data

WHY is it important to analyze multiple samples of data before making predictions?

Use the exercises below to help answer the Inquiry Question. Write the correct word or phrase on the lines provided.

1. Rewrite the question in your own words.

_____

_____

2. What key words do you see in the question?

_____

3. To _____ data means to study it and understand its meaning.

4. Write three synonyms for the word multiple. _____

5. When you make a _____ , you tell what you think will happen.

A shop sells three flavors of juice: apple, grape, and orange. The table shows which flavor customers chose in three different samples.

Flavor	Sample 1 (5 customers)	Sample 2 (15 customers)	Sample 3 (30 customers)
apple	0	4	11
grape	4	6	9
orange	1	5	10

6. Based on the results, what prediction can you make about the probability of a customer choosing apple juice in Sample 1? _____

7. Based on the results, what predction can you make about the probability of a customer choosing apple juice in Sample 3? _____

8. Which sample gives the most reliable results? _____

WHY is it important to analyze multiple samples of data before making predictions?

_____

_____

Lesson 3 Notetaking

Misleading Graphs and Statistics

Use Cornell notes to better understand the lesson's concepts. Complete each sentence by filling in the blanks with the correct word or phrase.

Questions	Notes
1. How do I identify a misleading graph?	Look at the scales on the _____ and _____ axes. To emphasize a change over time, there may have been a _____ of the scale interval on the _____ axis.
2. How do I identify misleading statistics?	Identify which _____ is being used to describe the set of data. Measures of center include the _____ , _____ , and _____ .

Summary

Describe at least two ways in which the display of data can influence the conclusions reached.

_____

_____

_____

_____

_____

Problem-Solving Investigation
Use a Graph

Case 3 Postage

The table shows the postage stamp rate from 1999 to 2009.

Make a graph of the data.

Predict the **year** the postage **rate** will reach **$0.52.**

Postage Stamp Rates	
Year	Cost ($)
1999	0.33
2001	0.34
2002	0.37
2006	0.39
2007	0.41
2008	0.42
2009	0.44

- Understand:

- Plan:

- Solve:

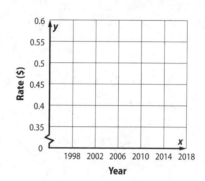

- Check:

Case 4 Trains

The lengths of various train rides are **4, 1, 2, 3, 6, 2, 3, 2, 5, 8, and 4 hours.**

Draw a box plot for the data set.

What percent of the train rides are **longer than 3 hours**?

- Understand:

- Plan:

- Solve:

- Check:

Inquiry Lab Guided Writing

Collect Data

HOW can you use the measures of center and the range to compare two populations?

Use the exercises below to help answer the Inquiry Question. Write the correct word or phrase on the lines provided.

1. Rewrite the question in your own words.

_____

_____

2. What key words do you see in the question?

_____

3. Mean, median, and mode are different measures of _____ for a set of numerical data.

4. _____ is the difference between the greatest and least data value.

5. _____ is the group from which samples are taken.

6. What are two populations you could compare at your school?

_____

_____

7. How could you collect numerical data from random samples of the populations?

_____

9. A graph shows a range of 5 for boys and a range of 3 for girls.

For which population is the date more varied? _____

HOW can you use the measures of center and the range to compare two populations?

_____

_____

_____

Lesson 4 Vocabulary
Compare Populations

Use the word cards to define each vocabulary word or phrase and give an example.

Word Cards

| double box plot | doble diagram de caja |

Definition

_____

_____

_____

Definición

_____

_____

_____

Example Sentence

_____

_____

- -

Word Cards

| double dot plot | doble diagram de puntos |

Definition

_____

_____

Definición

_____

_____

Example Sentence

_____

_____

Inquiry Lab Guided Writing

Visual Overlap of Data Distributions

WHAT does the ratio $\dfrac{\text{difference in means}}{\text{mean absolute deviation}}$ **tell you about**

how much visual overlap there is between two distributions

with similar variation?

Use the exercises below to help answer the Inquiry Question. Write the correct word or phrase on the lines provided.

1. Rewrite the question in your own words.

   _____

   _____

2. What key words do you see in the question?

   _____

3. _____ is a measure of center found by dividing the sum of numerical data by the number of items in the set of data.

4. The average of the absolute values of the differences between the data values

   and the mean is called the _____ .

5. _____ describes the spread of a set of data, or how scattered the data is.

6. When there is more visual overlap between distributions, the

   ratio _____ is smaller.
   There is less visual overlap as the ratio gets larger.

7. Which ratio shows more visual overlap: $\dfrac{12}{4.5}$ or $\dfrac{16}{4.5}$? _____

WHAT does the ratio $\dfrac{\text{difference in means}}{\text{mean absolute deviation}}$ tell you about how much visual

overlap there is between two distributions with similar variation?

_____

_____

_____

Lesson 5 Review Vocabulary

Select an Appropriate Display

Write the type of display best used for each situation. Choose from a bar graph, box plot, circle graph, double bar graph, histogram, line graph, or line plot.

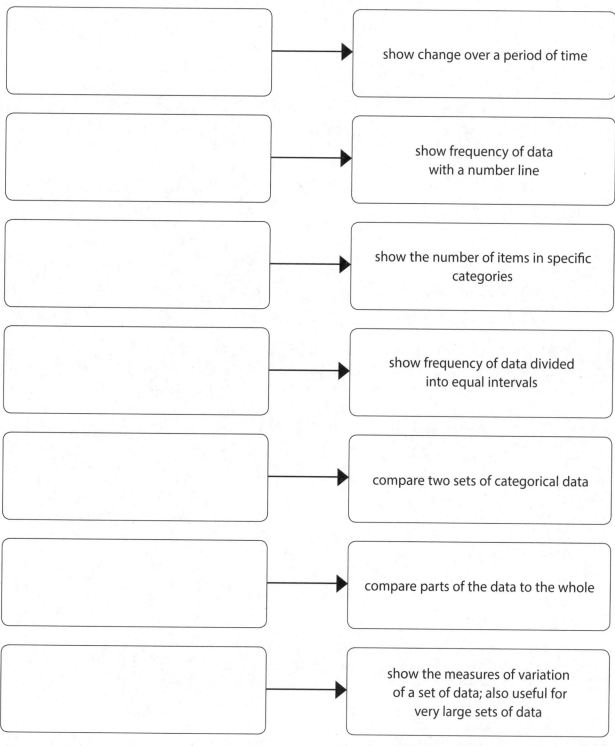

	show change over a period of time
	show frequency of data with a number line
	show the number of items in specific categories
	show frequency of data divided into equal intervals
	compare two sets of categorical data
	compare parts of the data to the whole
	show the measures of variation of a set of data; also useful for very large sets of data

What are VKVs® and How Do I Create Them?

Visual Kinethestic Vocabulary Cards® are flashcards that animate words by focusing on their structure, use, and meaning. The VKVs in this book are used to show cognates, or words that are similar in Spanish and English.

Step 1

Go to the back of your book to find the VKVs for the chapter vocabulary you are currently studying. Follow the cutting and folding instructions at the top of the page. The vocabulary word on the BLUE background is written in English. The Spanish word is on the ORANGE background.

Step 2

There are exercises for you to complete on the VKVs. When you understand the concept, you can complete each exercise. All exercises are written in English and Spanish. You only need to give the answer once.

Step 3

Individualize your VKV by writing notes, sketching diagrams, recording examples, and forming plurals (radius: radii or radiuses).

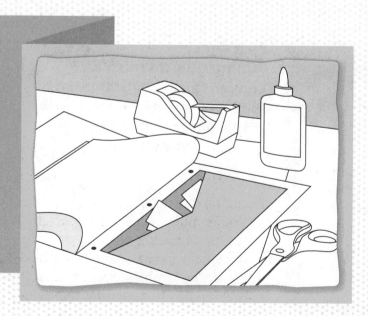

How Do I Store My VKVs?

Take a 6" x 9" envelope and cut away a V on one side only. Glue the envelope into the back cover of your book. Your VKVs can be stored in this pocket!

Remember you can use your VKVs ANY time in the school year to review new words in math, and add new information you learn. Why not create your own VKVs for other words you see and share them with others!

¿Qué son las VKV y cómo se crean?

Dinah Zike's
Visual
Kinesthetic
Vocabulary®

Las tarjetas de vocabulario visual y cinético (VKV) contienen palabras con animación que está basada en la estructura, uso y significado de las palabras. Las tarjetas de este libro sirven para mostrar cognados, que son palabras similares en español y en inglés.

Paso 1

Busca al final del libro las VKV que tienen el vocabulario del capítulo que estás estudiando. Sigue las instrucciones de cortar y doblar que se muestran al principio. La palabra de vocabulario con fondo AZUL está en inglés. La de español tiene fondo NARANJA.

Paso 2

Hay ejercicios para que completes con las VKV. Cuando entiendas el concepto, puedes completar cada ejercicio. Todos los ejercicios están escritos en inglés y español. Solo tienes que dar la respuesta una vez.

Paso 3

Da tu toque personal a las VKV escribiendo notas, haciendo diagramas, grabando ejemplos y formando plurales (radio: radios).

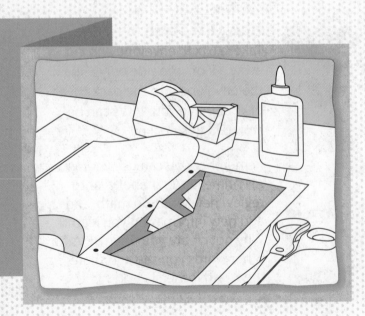

¿Cómo guardo mis VKV?

Corta en forma de "V" el lado de un sobre de 6" X 9". Pega el sobre en la contraportada de tu libro. Puedes guardar tus VKV en esos bolsillos. ¡Así de fácil!

Recuerda que puedes usar tus VKV en cualquier momento del año escolar para repasar nuevas palabras de matemáticas, y para añadir la nueva información. También puedes crear más VKV para otras palabras que veas, y poder compartirlas con los demás.

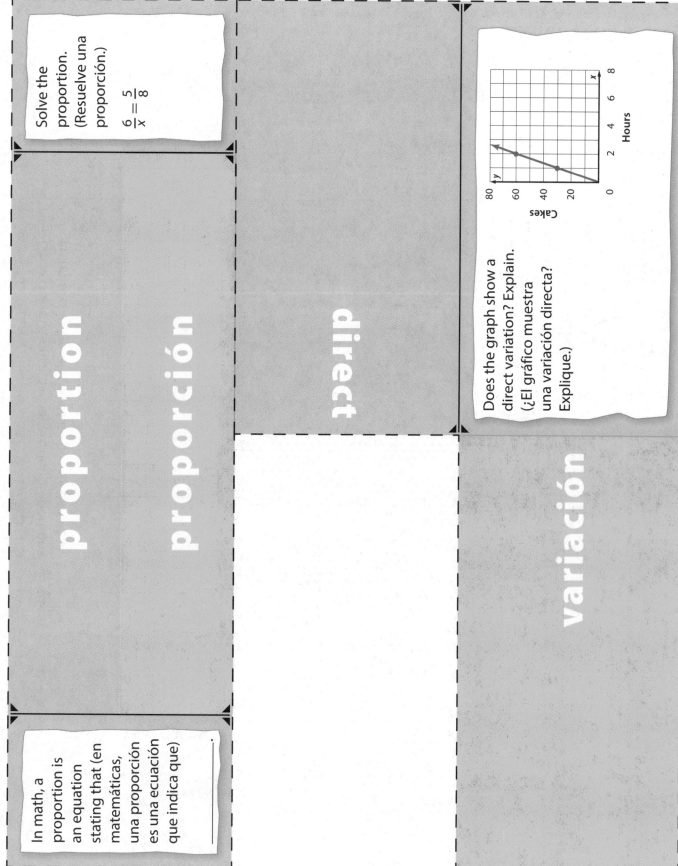

Solve the proportion. (Resuelve una proporción.)

$$\frac{6}{x} = \frac{5}{8}$$

proportion

proporción

direct

variación

Does the graph show a direct variation? Explain. (¿El gráfico muestra una variación directa? Explique.)

In math, a proportion is an equation stating that (en matemáticas, una proporción es una ecuación que indica que)

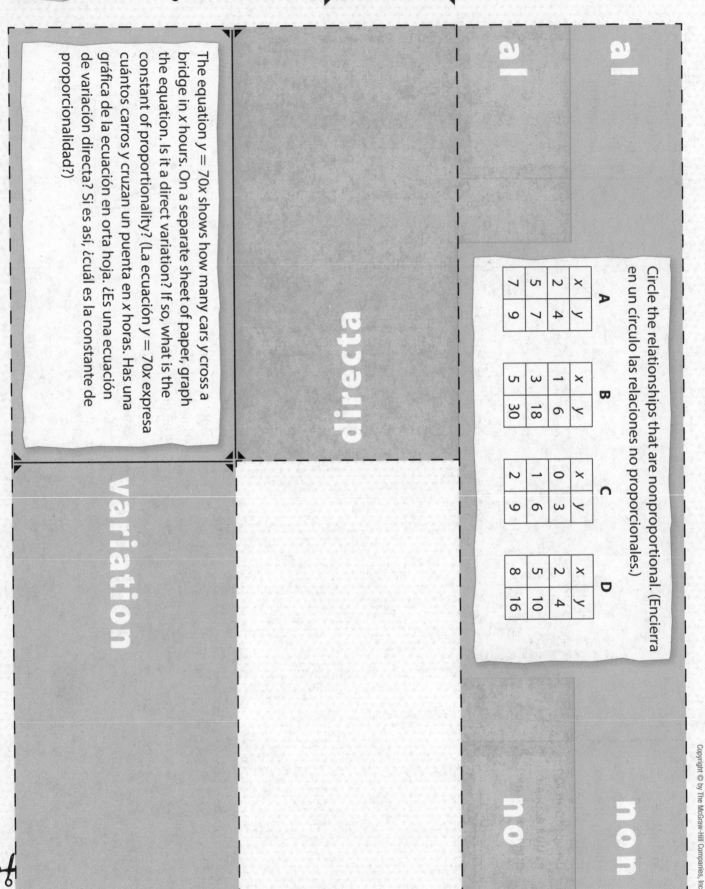

The equation $y = 70x$ shows how many cars y cross a bridge in x hours. On a separate sheet of paper, graph the equation. Is it a direct variation? If so, what is the constant of proportionality? (La ecuación $y = 70x$ expresa cuántos carros y cruzan un puente en x horas. Has una gráfica de la ecuación en orta hoja. ¿Es una ecuación de variación directa? Si es así, ¿cuál es la constante de variación directa? Si es así, ¿cuál es la constante de proporcionalidad?)

directa

variation

al

al

al

Circle the relationships that are nonproportional. (Encierra en un círculo las relaciones no proporcionales.)

A

x	y
2	4
5	7
7	9

B

x	y
1	6
3	18
5	30

C

x	y
0	3
1	6
2	9

D

x	y
2	5
4	10
8	16

non

no

Dinah Zike's
V K V
**Visual
Kinesthetic
Vocabulary**®

Chapter 2

cut on all dashed lines

fold on all solid lines

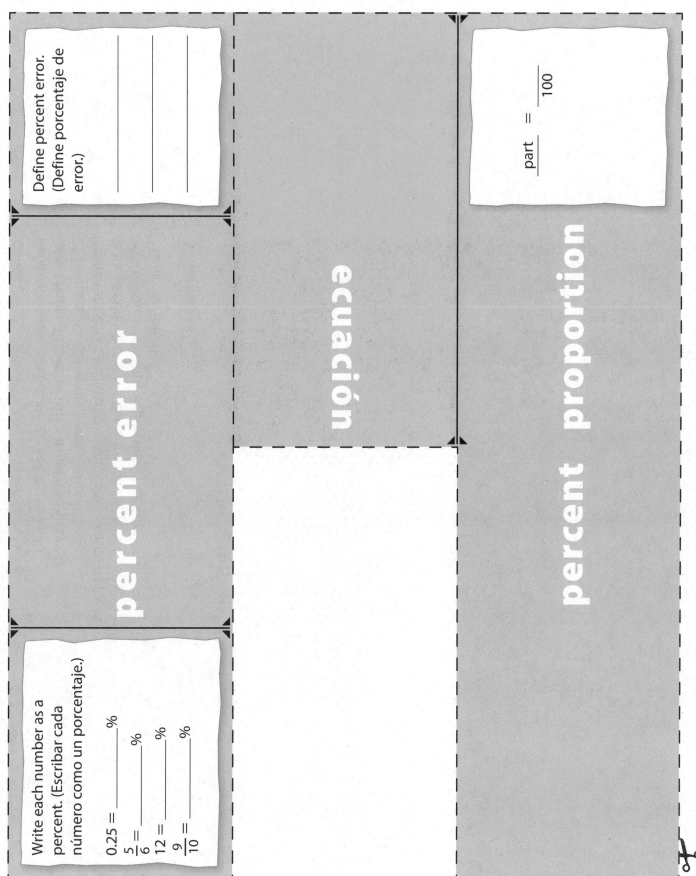

Define percent error.
(Define porcentaje de error.)

$$\frac{part}{} = \frac{}{100}$$

percent error

ecuación

percent proportion

Write each number as a percent. (Escribar cada número como un porcentaje.)

$0.25 =$ _____ %

$\frac{5}{6} =$ _____ %

$12 =$ _____ %

$\frac{9}{10} =$ _____ %

Dinah Zike's
Visual
Kinesthetic
Vocabulary®

Chapter 2

✂ cut on all dashed lines

📄 fold on all solid lines

proporción porcentual

equation

part = ————
        •

de error

porcentaje

Sofia estimated that a drive to the beach would take 3 hours. The actual drive lasted for 3 hours and 25 minutes. What was the percent error of Sofia's estimate? (Sofía calculó que un recorrido en auto hasta la playa tardaría 3 horas. El recorrido real tardó 3 horas y 25 minutos. ¿Cual fue el porcentaje de error en el cálculo de Sofía?)

Answer: about ———— %

Dinah Zike's
Visual
Kinesthetic
Vocabulary®

Chapter 3

✂ cut on all dashed lines

fold on all solid lines

Write the additive inverse of each number below. (Escribe el inverso aditivo de los siguientes números.)

13 _____ _____

_____ −2

−25 _____

_____ 1

additive inverse

absolute value

Define additive inverse. (Define inverso aditivo.)

_____
_____
_____
_____

Define absolute value. (Define valor absoluto.)

_____
_____
_____
_____
_____

Dinah Zike's
**Visual
Kinesthetic
Vocabulary**®

Chapter 3

✂ cut on all dashed lines

📷 fold on all solid lines

aditivo

valor absoluto

inverso

Circle the word that has the same meaning as *additive inverse*. (Encierra en un círculo la palabra que significa lo mismo que inverso aditivo.)

negative integer   opposite

number line   absolute value

Evaluate each expression. (Evalúa cada expresión.)

|−23| = _____ |8| = _____

|−7| + |10| = _____ |6| − |−6| = _____

Dinah Zike's
Visual
Kinesthetic
Vocabulary ®

✂ cut on all dashed lines   ⬜ fold on all solid lines

Write the opposite of each number below. (Escribe el opuesto de los siguientes números.)

12 _____

−54 _____

−16 _____

25 _____

Explain how to graph an integer on a number line. (Explica cómo representar gráficamente un entero en una recta numérica.)

_____

opposites

graph

ficar

uestos

Graph –6, 2, 8, and –3 on the number line. (Representa gráficamente –6, 2, 8, y –3 en la recta numérica.)

-8 -6 -4 -2 0 2 4 6 8 10 12

Explain why 6 and –6 are opposites. (Explica por qué 6 y –6 son opuestos.)

Why is bar notation used to represent repeating decimals? (¿Por qué la notación de barras se utiliza para representar números decimales periódicos?)

bar notation

common denominator

What does the bar in 0.1$\overline{6}$ mean? (¿Qué significa la barra sobre 0.1$\overline{6}$?)

How does writing fractions with a common denominator help you compare them? (¿De qué forma escribir fracciones con un común denominador te ayuda a compararlas entre sí?)

Dinah Zike's
Visual
Kinesthetic
Vocabulary®

Chapter 4

cut on all dashed lines

fold on all solid lines

común denominador

de barra

notación

Use bar notation to rewrite each decimal. (Vuelve a escribir cada número decimal con notación de barras.)

$0.7777\ldots =$ _____

$9.3555\ldots =$ _____

$-0.337337\ldots =$ _____

Rewrite $\frac{1}{8}$ and $\frac{1}{5}$ with a common denominator.

(Escribe $\frac{1}{8}$ y $\frac{1}{5}$ de manera que ambos tengan un común denominador.)

$\frac{1}{5} =$ _____

$\frac{1}{8} =$ _____

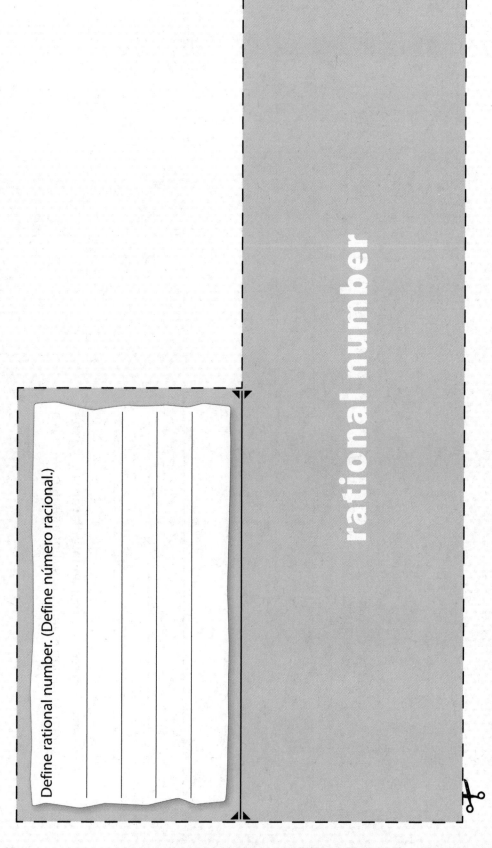

rational number

Define rational number. (Define número racional.)

Dinah Zike's
V K V
Visual
Kinesthetic
Vocabulary®

Chapter 4

✂ cut on all dashed lines

▭ fold on all solid lines

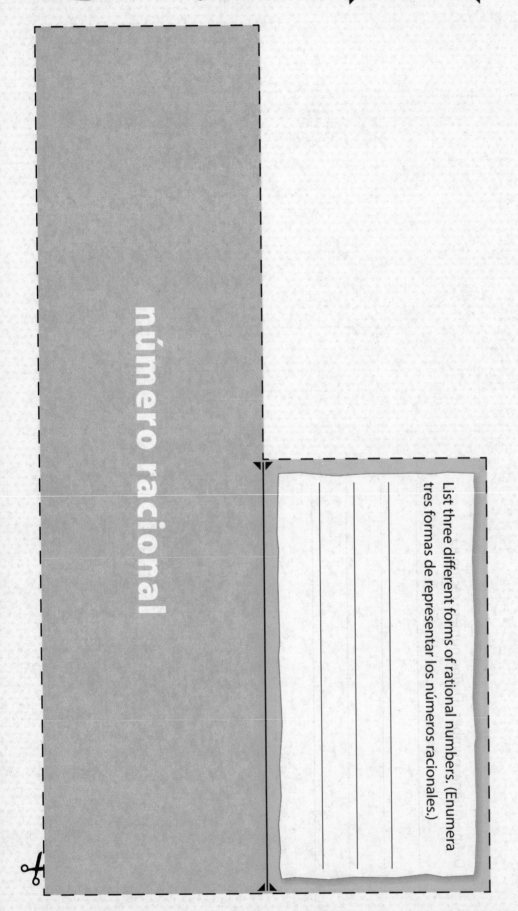

número racional

List three different forms of rational numbers. (Enumera tres formas de representar los números racionales.)

Dinah Zike's
Visual
Kinesthetic
Vocabulary®

Chapter 5

✂ cut on all dashed lines

fold on all solid lines

coefficient

Define coefficient. (Define coeficiente.)

What property is used when factoring a linear expression? (¿Qué propiedad se utiliza para factorizar una función lineal?)

factored form

Rewrite 9x − 15y in factored form. (Escribe la forma factorizada de la función 9x − 15y.)

Circle the greatest common factor of 15x and 27xy. (Encierra en un círculo el máximo común divisor de 15x y 27xy.)

3	y
3x	
5	9
x	

factor

Dinah Zike's
Visual Kinesthetic Vocabulary ®

✂ cut on all dashed lines

fold on all solid lines

iciente

factorizada

izar

Write about a time when writing a linear expression in factored form might be useful. (Escribe sobre una situación en la que sería útil escribir la forma factorizada de una función lineal.)

Circle the coefficient in each expression. (Encierra en un círculo el coeficiente de cada expresión.)

$12 - 5x$ $19m - 24$

$9p + 17r$ $37 + 11a$

Factor each expression. (Factoriza cada expresión.)

$12x + 3 =$ _____

$27a + 12b =$ _____

$16m - 18 =$ _____

forma

✂ cut on all dashed lines   fold on all solid lines

Find the GCF of the pair of monomials. (Halla el MCD de ambos monomios.)

6x, 21xy

_____

expression

monomial

lineal

Is $a^2 - 6$ a linear expression? Explain. (¿Es la expresión $a^2 - 6$ una función lineal? Explica.)

✂ cut on all dashed lines   ▭ fold on all solid lines

Dinah Zike's
Visual Kinesthetic Vocabulary®

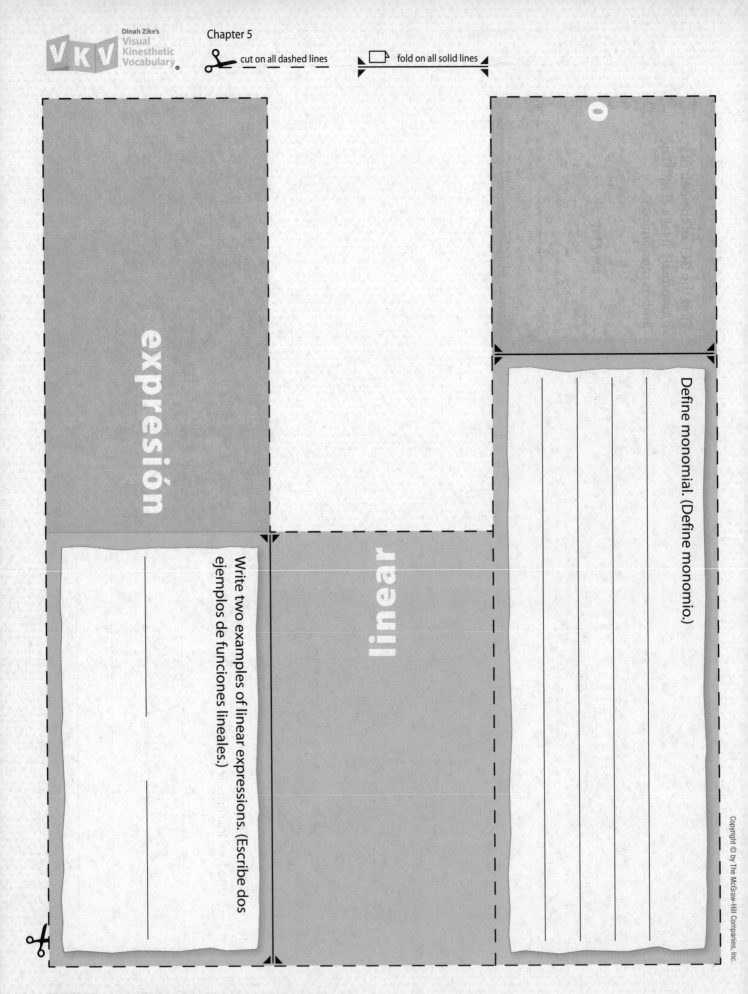

expresión

linear

Define monomial. (Define monomio.)

Write two examples of linear expressions. (Escribe dos ejemplos de funciones lineales.)

Dinah Zike's
VKV Visual
Kinesthetic
Vocabulary®

Chapter 6

✂ cut on all dashed lines   fold on all solid lines

What value of x makes the equation true? (¿Qué valor de x hace la ecuación verdadera?)

_____

$x + 8 = 13$

solution

Circle the equations that are equivalent to $x = 9$. (Encierra en un círculo las ecuaciones equivalentes a $x = 9$.)

$x - 5 = 4$ $3 + x = 12$

$x + 12 = 20$

equivalent equation

Dinah Zike's
Visual
Kinesthetic
Vocabulary®

ción

ecuación equivalente

Find the solution of each equation. (Halla la solución de cada ecuación.)

$x - 12 = 8$ _____

$9 + y = 17$ _____

Are the equations $x + 11 = 14$ and $x = 3$ equivalent? Explain. (¿Son las ecuaciones $x + 11 = 14$ y $x = 3$ equivalentes? Explica.)

_____
_____
_____
_____

Chapter 7

Dinah Zike's
VKV Visual
Kinesthetic
Vocabulary®

✂ cut on all dashed lines   ⬜ fold on all solid lines

How many planes intersect to form a cube?
(¿Cuantos planos se intersecan en un cubo?)

ilindro

Define congruent. (Define congruente.)

Cones and cylinders are not polyhedrons. Explain why. (Explica por qué los conos y los cilindros no son poliedros.)

plane

cone

congruent

Dinah Zike's
Visual
Kinesthetic
Vocabulary®

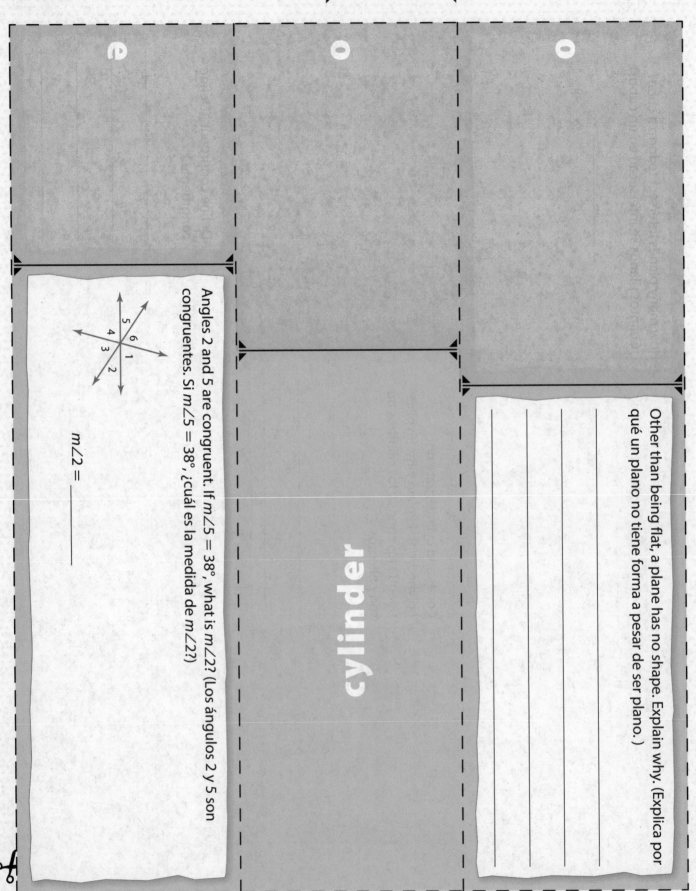

e

o

o

Angles 2 and 5 are congruent. If $m\angle 5 = 38°$, what is $m\angle 2$? (Los ángulos 2 y 5 son congruentes. Si $m\angle 5 = 38°$, ¿cuál es la medida de $m\angle 2$?)

$m\angle 2 =$ _____

cylinder

Other than being flat, a plane has no shape. Explain why. (Explica por qué un plano no tiene forma a pesar de ser plano.)

scale model

factor de

What is the scale factor of a dollhouse if 5 centimeters represents 1 meter? (¿Cuál es el factor de escala de una casa de muñecas en la que 5 centímetros representan un metro?)

modelo a escala

factor

List three examples where you might find scale models used. (Menciona tres situaciones en las que se utilizan modelos a escala.)

Dinah Zike's
VKV
Visual
Kinesthetic
Vocabulary®

✂ cut on all dashed lines   ▱ fold on all solid lines

center

nferencia

circle

Define center. (Define centro.)

Define circle. (Define círculo.)

Dinah Zike's
**Visual
Kinesthetic
Vocabulary**®

Chapter 8

✂ cut on all dashed lines

▢ fold on all solid lines

circumference

círculo

Use *center* to describe the radius and diameter of a circle. (Utiliza la palabra *centro* para describir el radio y el diámetro de un círculo.)

ro

Find the circumference. (Halla la circunferencia.)

15 m

Dinah Zike's
VKV
Visual
Kinesthetic
Vocabulary®

Chapter 8

✂ cut on all dashed lines

✂ fold on all solid lines

A circle's diameter is _____ the length of the circle's radius. (El diámetro de un círculo es el _____ de la longitud del radio del círculo.)

If you know the length of a circle's radius, what three other measurements can you find? (¿Cuáles tres medidas puedes calcular con la longitud del radio de un círculo?)

Describe a real-world composite figure, explaining the object's purpose, and the figures of which it is composed. (Describe un objeto real cuya forma sea una figura compuesta. Explica su función y menciona las figuras de las cuales se compone.)

Define composite figure. (Define figura compuesto.)

diameter

radius

composite figure

compuesto

o

ámetro

Use two polygons to draw a composite figure. (Dibuja una figura compuesta por dos polígonos.)

figura

Draw a radius of the circle. (Dibuja el radio del círculo.)

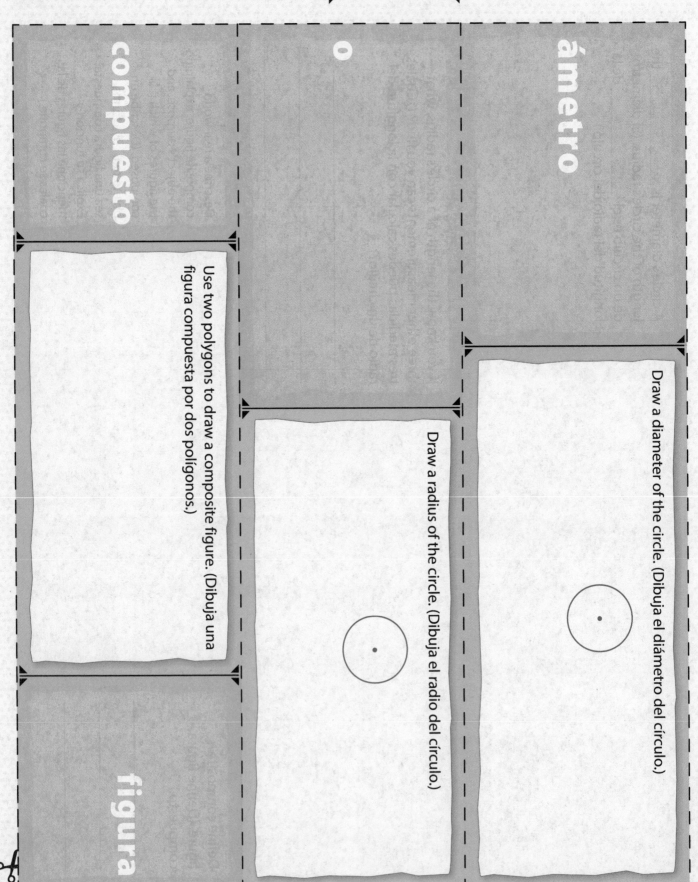

Draw a diameter of the circle. (Dibuja el diámetro del círculo.)

Dinah Zike's
Visual
Kinesthetic
Vocabulary®

Chapter 8

cut on all dashed lines

fold on all solid lines

The formula for the area of a circle is $A = \pi r^2$. Write the formula for the area of a semicircle.
(La fórmula para calcular el área de un círculo es $A = \pi r^2$. Escribe la fórmula para calcular el área de un semicírculo.)

$A = $ _____

semicircle

lateral surface area

Define lateral surface area. (Define área de superficie lateral.)

Dinah Zike's
V K V
Visual
Kinesthetic
Vocabulary ®

Chapter 8

✂ cut on all dashed lines

▭ fold on all solid lines

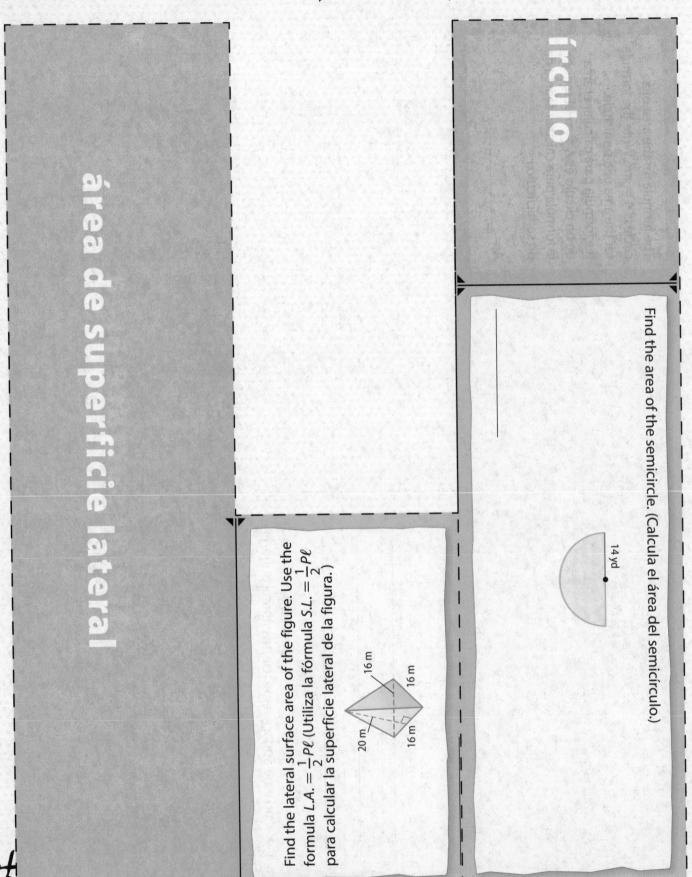

írculo

Find the area of the semicircle. (Calcula el área del semicírculo.)

14 yd

Find the lateral surface area of the figure. Use the formula $L.A. = \frac{1}{2}P\ell$ (Utiliza la fórmula $S.L. = \frac{1}{2}P\ell$ para calcular la superficie lateral de la figura.)

16 m
16 m
20 m
16 m

área de superficie lateral

✂ cut on all dashed lines

📱 fold on all solid lines

regular pyramid

The base of a regular pyramid is a regular _____.
(La base de una pirámide regular es un _____.)

Dinah Zike's
Visual Kinesthetic Vocabulary ®

cut on all dashed lines

fold on all solid lines

pirámide regular

Define regular pyramid. (Define pirámide regular.)

Dinah Zike's
**Visual
Kinesthetic
Vocabulary** ®

✂ ---- cut on all dashed lines

◁ 📄 ▷ fold on all solid lines

Define simulation. (Define simulación.)

Define permutation. (Define permutación.)

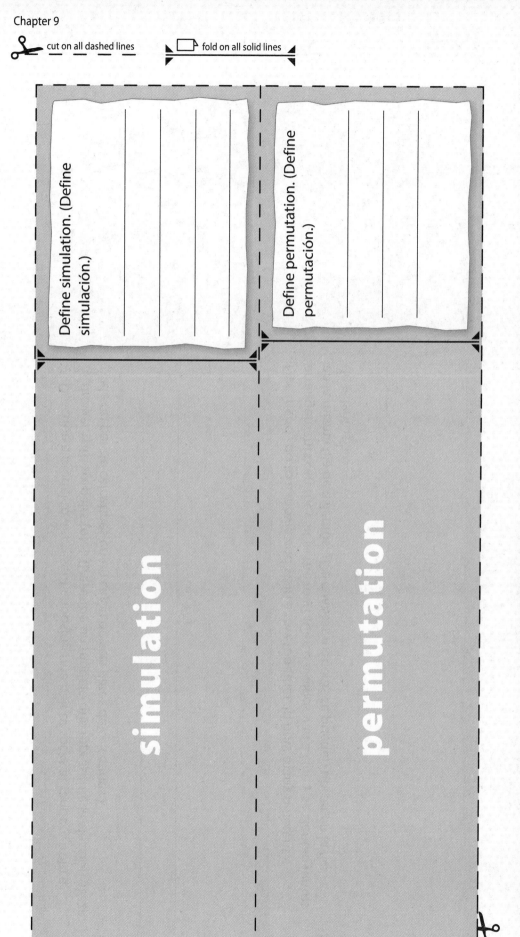

simulation

permutation

✂ cut on all dashed lines   fold on all solid lines

ción

ción

Describe a model to represent choosing one pair of socks at random from a drawer with six pairs total. (Describe un modelo que represente la elección al azar de un par de calcetines en un cajón de seis pares de calcetines.)

Five friends go to a movie and sit in a row together. In how many different arrangements can they sit in the row? (Cinco amigos van a cine y se sientan juntos en la misma fila de asientos. ¿De cuántas maneras se pueden ubicar en la fila de asientos?)

Dinah Zike's
VKV
Visual
Kinesthetic
Vocabulary ®

✂ cut on all dashed lines

⬜ fold on all solid lines

theoretical probability

experimental

What is the theoretical probability of rolling a 3 with a number cube? (¿Cuál es la probabilidad teórica de obtener un 3 al lanzar un dado?)

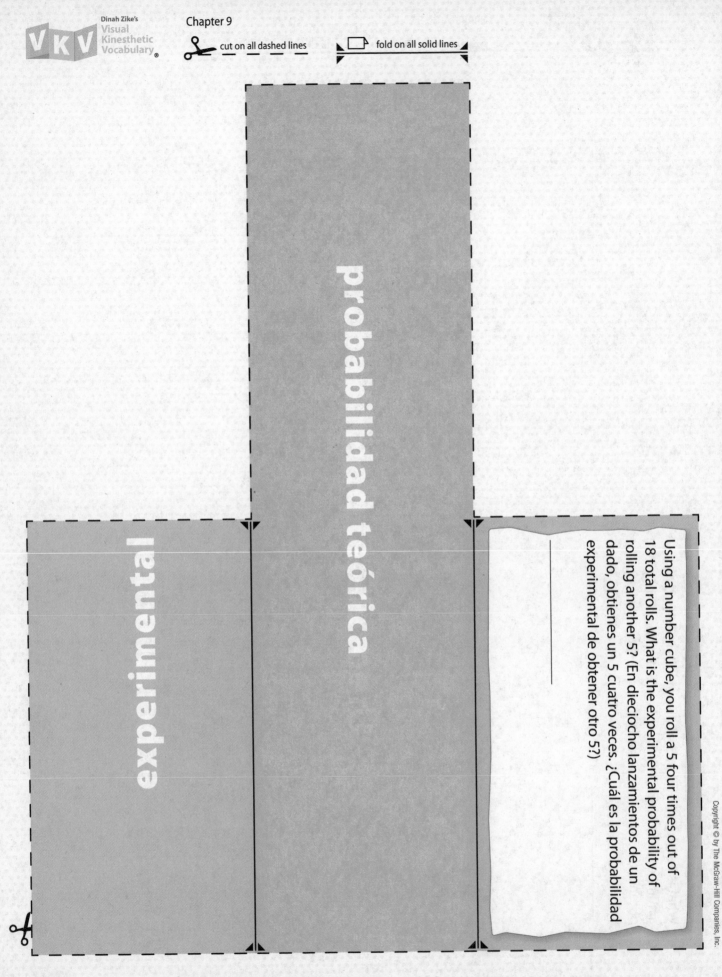

probabilidad teórica

experimental

Using a number cube, you roll a 5 four times out of 18 total rolls. What is the experimental probability of rolling another 5? (En dieciocho lanzamientos de un dado, obtienes un 5 cuatro veces. ¿Cuál es la probabilidad experimental de obtener otro 5?)

✂ cut on all dashed lines   fold on all solid lines

independent events

dependientes

Give an example of a dependent event. (Da un ejemplo de un suceso dependiente.)

Dinah Zike's
**Visual
Kinesthetic
Vocabulary** ®

✂ cut on all dashed lines   fold on all solid lines

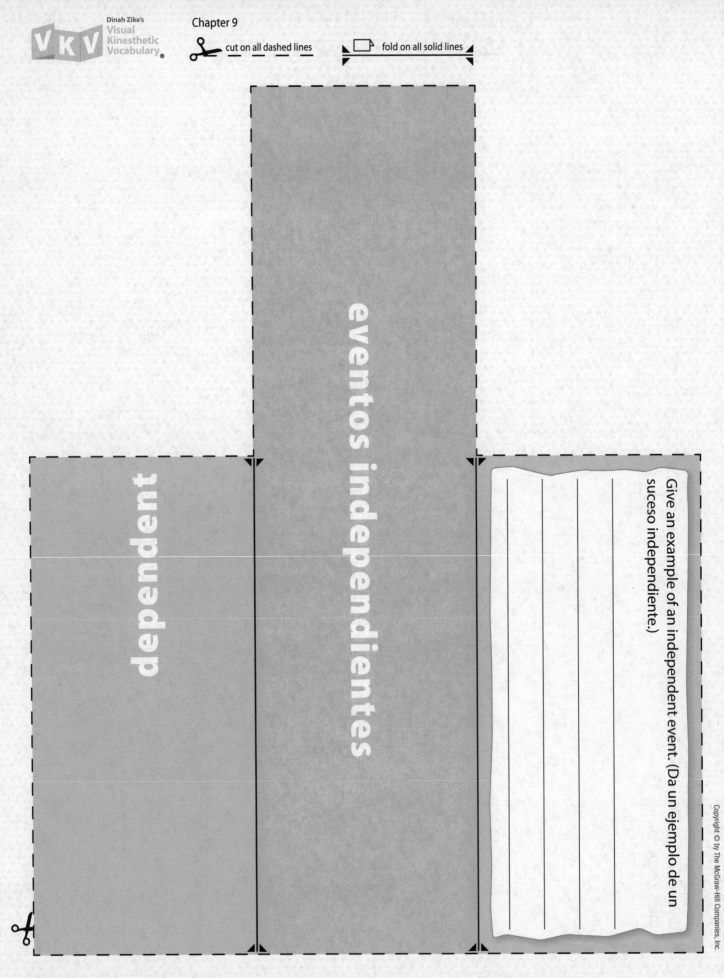

eventos independientes

dependent

Give an example of an independent event. (Da un ejemplo de un suceso independiente.)

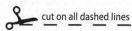

cut on all dashed lines   fold on all solid lines

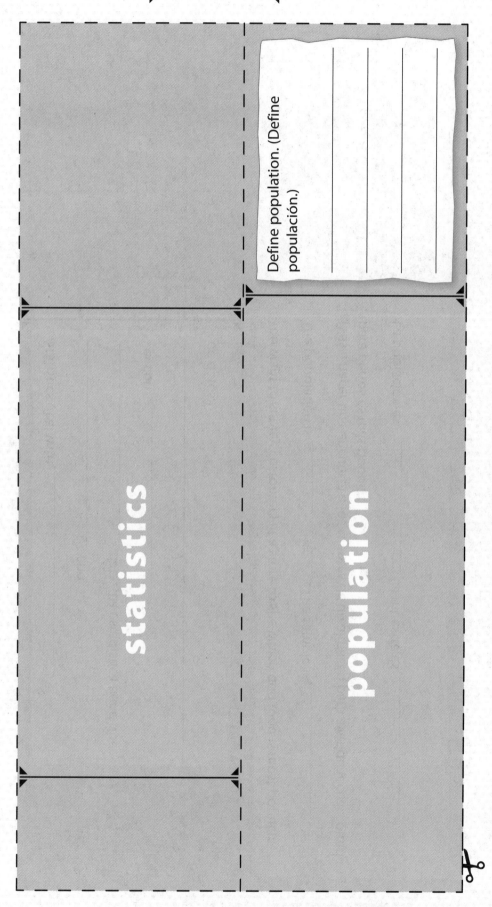

Define population. (Define población.)

statistics

population

Dinah Zike's
VKV
Visual
Kinesthetic
Vocabulary ®

Chapter 10

✂ cut on all dashed lines

🗇 fold on all solid lines

ción

dística

Statistics deal with _____, and

_____ data. (La estadística tiene que

ver con _____ y

_____

de datos.)

You want to survey customers at a store to see which dog food is most popular.

The population is _____. The sample is _____.

(Vas a hacer una encuesta a los clientes de un almacén para averiguar cuál comida

para perros es más popular.

La población es _____

_____. La muestra es

_____.)

e